Lecture Notes in Computer Science 5425

Commenced Publication in 1973
Founding and Former Series Editors:
Gerhard Goos, Juris Hartmanis, and Jan van Leeuwen

Editorial Board

Eitan Altman   Augustin Chaintreau (Eds.)

Network Control and Optimization

Second EuroFGI Workshop, NET-COOP 2008
Paris, France, September 8-10, 2008
Revised Selected Papers

 Springer

Volume Editors

Eitan Altman
INRIA
2004 Route des Lucioles, 06902 Sophia-Antipolis CEDEX, France
E-mail: eitan.altman@sophia.inria.fr

Augustin Chaintreau
Thomson
46 quai Alphonse le Gallo, 92648 Boulogne CEDEX, France
E-mail: augustin.chaintreau@thomson.net

Library of Congress Control Number: Applied for

CR Subject Classification (1998): C.2.4, F.2, D.2, D.1.3, H.4

LNCS Sublibrary: SL 5 – Computer Communication Networks
and Telecommunications

ISSN   0302-9743
ISBN-10   3-642-00392-3 Springer Berlin Heidelberg New York
ISBN-13   978-3-642-00392-9 Springer Berlin Heidelberg New York

springer.com

© Springer-Verlag Berlin Heidelberg 2009
Printed in Germany

Typesetting: Camera-ready by author, data conversion by Scientific Publishing Services, Chennai, India
Printed on acid-free paper   SPIN: 12620258   06/3180   5 4 3 2 1 0

Preface

We are pleased to present the proceedings of the Second Workshop on Network Control and Optimization (NET-COOP) that took place during September 8-10, 2008, in Paris, France, sponsored by the EuroFGI Network of Excellence.

Thanks to the generous support of EuroFGI, we were able to combine contributed sessions with an outstanding list of invited plenary speakers: Thomas Bonald, Vivek S. Borkar, Pierre Fraigniaud, Ayavaldi J. Ganesh, Takis Konstantopoulos, Alexandre Proutière, R. Srikant, Sacha Stolyar, Laurent Viennot, Milan Vojnovic, Damon Wischik.

The workshop received 27 submissions which underwent a thorough review process. The Program Committee selected 13 of these as contributed papers. The proceedings include in addition two invited papers in connection with two of the plenary talks.

Continuing a young tradition, the workshop aims at developing research on control and optimization of the Internet, ranging from performance evaluation and optimization of general stochastic networks to more specific targets such as lower-layer functionalities in mobile networks, routing for computational grids, game theoretic approaches to access control, cooperation, competition and adversary capacities in diverse environments.

November 2008   Eitan Altman
                                                    Augustin Chaintreau

Organization

NET-COOP 2008 was co-organized by Thomson and TELECOM SudParis, France

Organizing Committee

General Chair: Laurent Massoulié (Thomson, France)
General Co-chair: Tijani Chahed (TELECOM SudParis, France)
Program Co-chairs: Eitan Altman (INRIA, France) and Augustin
                             Chaintreau (Thomson, France)
Web Chair: Nelson Vicuna (Avignon University, France)

Program Committee

Ivo Adan, TU/e
Panayotis Antoniadis, U. Paris 6
Konstantin Avratchenkov, INRIA
Urtzi Ayesta, CNRS
Tamer Basar, UIUC
Randall Berry, Northwestern
Thomas Bonald, OrangeLabs
Vivek S. Borkar, TIFR
Onno Boxma, TU Eindhoven
Costas Courcoubetis, AUEB
Bruno Gaujal, INRIA
Moshe Haviv, HUJI
Nidhi Hedge, OrangeLabs
Mikael Johansson, KTH
Hisao Kameda, U. Tsukuba
Peter Key, Microsoft Res.
Anurag Kumar, IISc
Nikolaos Laoutaris, Telefonica
Marc Lelarge, INRIA-ENS
Steven Low, CIT
D. Manjunath, IIT-Mumbai

Ravi Mazumdar, U. Waterloo
Pietro Michiardi, Eurecom
Giovanni Neglia, INRIA
José Niño-Mora, U. Carlos III
    de Madrid
Fernando Paganini, U. ORT
Pascale Primet, ENS-Lyon
Alexandre Proutière, Microsoft Res.
Guy Pujolle, U. Paris 6
Jacques Resing, TU Eindhoven
Ulrich Rieder, U. Ulm
Nahum Shimkin, Technion Haifa
George Stamoulis, AUEB
Nicolas Stier, Columbia
Leandros Tassiulas, U. Thessaly
Corinne Touati, INRIA
Don Towsley, UMASS
Darryl Veitch, U. Melbourne
Jorma Virtamo, HUT
Milan Vojnovic, Microsoft Res.
Bert Zwart, GeorgiaTech

Sponsoring Institutions

Euro-NF
INRIA
Thomson

Table of Contents

TCP and Congestion Control

Wireless Networks

A Comparison of Bilateral and Multilateral Exchanges for Peer-Assisted Content Distribution

Christina Aperjis, Michael J. Freedman, and Ramesh Johari

Abstract. Peer-assisted content distribution matches user demand for content with available supply at other peers in the network. Inspired by this supply-and-demand interpretation of the nature of content sharing, we employ *price theory* to study peer-assisted content distribution. In this approach, the market-clearing prices are those which exactly align supply and demand, and the system is studied through the characterization of price equilibria. We rigorously analyze the efficiency and robustness gains that are enabled by price-based multilateral exchange. We show that multilateral exchanges satisfy several desirable efficiency and robustness properties that bilateral exchanges do not, *e.g.*, equilibria in bilateral exchange may fail to exist, be inefficient if they do exist, and fail to remain robust to collusive deviations even if they are Pareto efficient. Further, we show that an equilibrium in bilateral exchange corresponds to a multilateral exchange equilibrium if and only if it is robust to deviations by coalitions of users.

1 Introduction

Peer-to-peer systems have been wildly successful as a disruptive technology for content distribution. Varying accounts place peer-to-peer (P2P) traffic as comprising anywhere between 35% and 90% of "all" Internet traffic [1]. Early P2P systems did not provide any incentives for participation, leading to extensive freeloading [2, 3]. The P2P community responded with mechanisms to prevent freeloading by incentivizing sharing on a *bilateral barter* basis, as used by BitTorrent [4] and its variants [5, 6], where peers can achieve better download performance from peers to which they are simultaneously uploading.

While BitTorrent's usage numbers are certainly impressive, it can only perform bilateral barter by matching up well-suited pairs of nodes that have disjoint subsets of a file (or, more generally, files), and it is often hard to find good reciprocation with bilateral barter alone. Furthermore, potential "free-riding" attacks have been observed [7, 8, 9, 5], and altruistic uploading often turns out to be critical for providing continued content availability [10].

Another alternative is to use *market-based multilateral exchange* to match user demand for content to available supply at other peers in the system. This approach uses virtual currency and assigns a budget to each peer, which decreases when downloading and increases when uploading. Monetary incentives in a virtual currency have been previously proposed to incentivize uploading in P2P systems [11, 12, 13, 14]. In this paper, we compare bilateral and multilateral exchanges for peer-assisted content distribution.

We provide a formal comparison of P2P system designs with bilateral barter, such as BitTorrent, and a market-based exchange of content enabled by a price mechanism to match supply and demand. We start in Section §2 with a fundamental abstraction

E. Altman and A. Chaintreau (Eds.): NET-COOP 2008, LNCS 5425, pp. 1–8, 2009.

of content exchange in systems with bilateral barter: *exchange ratios*. The exchange ratio from one peer to another gives the download rate received per unit upload rate. Exchange ratios are a useful formal tool because they directly allow us to compare bilateral P2P systems with price-based multilateral P2P systems.

In §3 and §4, we compare bilateral and multilateral P2P systems through the allocations that arise at equilibria. In particular, we show that a multilateral price-based exchange scheme satisfies a number of desirable properties lacking in bilateral exchange, *e.g.*, equilibria in bilateral exchange may fail to exist, be inefficient if they do exist, and fail to remain robust to collusive deviations even if they exist and are efficient. We show that with an additional technical condition, a bilateral equilibrium corresponds to a multilateral equilibrium if and only if it is robust to deviations by coalitions of users.

2   Exchange Ratios in Bilateral Protocols

The BitTorrent protocol and its variants enable exchange on a *bilateral* basis between peers: a peer i uploads to peer j if and only if peer j uploads to peer i in return. While such protocols are traditionally studied solely through the rates that peers obtain, in this section we provide an interpretation of these protocols through *exchange ratios*. As exchange ratios can be interpreted in terms of prices, these ratios will allow us to compare bilateral and multilateral P2P systems in the following section.

Let r_{ij} denote the rate sent from peer i to peer j in an instantiation of a BitTorrent swarm. We define the *exchange ratio* between peer i and peer j as the ratio $\gamma_{ij} = r_{ji}/r_{ij}$; this is the download rate received by i from j, per unit of rate uploaded to j. By definition, $\gamma_{ij} = 1/\gamma_{ji}$. Clearly, a rational peer i would prefer to download from peers with which he has higher exchange ratios.

The exchange ratio has a natural interpretation in terms of prices. An equivalent story emerges if we assume that peers charge each other for content in a common monetary unit, but that all transactions are *settlement-free*, *i.e.*, no money ever changes hands. In this case, if peer i charged peer j a price p_{ij} per unit rate, the exchange of content between peers i and j must satisfy $p_{ij}r_{ij} = p_{ji}r_{ji}$. Note that the preceding condition thus shows the exchange ratio is equivalent to the ratio of bilateral prices: $\gamma_{ij} = p_{ij}/p_{ji}$ (as long as the prices and rates are nonzero). The rates achieved by the BitTorrent and BitTyrant [5] protocols can be naturally modeled through exchange ratios [15].

The preceding discussion highlights the fact that the rates in a bilateral P2P system can be interpreted via exchange ratios. Thus far we have assumed that *transfer rates* are given, and exchange ratios are computed from these rates. In the next section, we turn this relationship around: we explicitly consider an abstraction of bilateral P2P systems where peers react to given exchange ratios, and compare the resulting outcomes to price-based multilateral exchange.

3   Bilateral and Multilateral Equilibria

Motivated by the discussion in the preceding section, this section rigorously analyzes the efficiency properties of price-based bilateral and multilateral mechanisms. Assuming that peers explicitly react to exchange ratios or prices, we compare the schemes through their resulting price equilibria.

In the formal model we consider, a set of peers N shares a set of files F. Peer i has a subset of the files $F_i \subseteq F$, and is interested in downloading files in $T_i \subseteq F - F_i$. We use

Bilateral Peer Optimization:	**Multilateral Peer Optimization:**
maximize $V_i(\mathsf{d}_i)$	maximize $V_i(\mathsf{d}_i)$
subject to $d_{if} = \sum_j r_{jif}$ for all f	subject to $d_{if} = \sum_j r_{jif}$ for all f
$\qquad r_{ijf} = 0$ if $f \notin F_i$	$\qquad r_{ijf} = 0$ if $f \notin F_i$
$\qquad \sum_{j,f} r_{ijf} \leq B_i$	$\qquad \sum_{j,f} r_{ijf} \leq B_i$
$\qquad \sum_f r_{jif} = \gamma_{ij} \sum_f r_{ijf}$ for all j	$\qquad \sum_{j,f} p_j r_{jif} = p_i \sum_{j,f} r_{ijf}$
$\qquad r \geq 0.$	$\qquad r \geq 0.$

Fig. 1. Optimization problems for price-based exchange

r_{ijf} to denote the rate at which user i uploads file f to user j. We then let $d_{if} = \sum_j r_{jif}$ be the total rate at which user i downloads file f. We use sans serif to denote vectors, *e.g.*, $\mathsf{d}_i = (d_{if}, f \in T_i)$ is the vector of download rates for user i. We measure the desirability of a download vector to peer i by a *utility function* $V_i(\mathsf{d}_i)$ that is nondecreasing in every d_{if} for $f \in T_i$. We ignore any resource constraints within the network; we assume that transfers are only constrained by the upload capacities of peers. The upload capacity of peer i is denoted B_i.

We start by considering peers' behavior in bilateral schemes, given a vector of exchange ratios (γ_{ij}). Peer i solves the bilateral optimization problem given in Figure 1. Note that we allow peers to bilaterally exchange content over multiple files, as the more general design, even though this is not typically supported by swarming systems like BitTorrent. This more general design makes it possible to explicitly identify the relative demand for files and reward peers that share more popular content.

By contrast, in a *multilateral price-based exchange*, the system maintains one price per peer, and peers optimize with respect to these prices. In a slight abuse of notation, we denote the price of a peer i by p_i. Figure 1 also gives the peer optimization problem in multilateral price-based exchange. Note that the first three constraints (giving download rates, ensuring peers only upload files they possess, and meeting the upload capacity constraint) are identical to the bilateral peer optimization. While the bilateral exchange implicitly requires peer i to download only from those peers to whom he uploads, no such constraint is imposed on multilateral exchanges: peer i accrues capital for uploading, and he can spend this capital however he wishes for downloading.

For bilateral (resp., multilateral) exchange, an *equilibrium* is a combination of a rate allocation vector and an exchange ratio vector (resp., price vector) such that all peers have solved their corresponding optimization problems. In this case, the exchange ratios (resp., prices) have exactly aligned supply and demand: for any i, j, f, the transfer rate r_{ijf} is simultaneously an optimal choice for both the uploader i and downloader j.

Definition 1. *The rate allocation* r^* *and the exchange ratios* $(\gamma_{ij}^*, i, j \in N)$ *with* $\gamma_{ij}^* > 0$ *for all* $i, j \in N$ *constitute a **bilateral equilibrium** if for each peer i, r^* solves the Bilateral Peer Optimization problem given exchange ratios* $(\gamma_{ij}^*, j \in N)$.

Definition 2. *The rate allocation* r^* *and the peer prices* $(p_i^*, i \in N)$ *with* $p_i^* > 0$ *for all* $i \in N$ *constitute a **multilateral equilibrium** if for each peer j, r^* solves the Multilateral Peer Optimization problem given prices* $(p_i^*, i \in N)$.

This latter is the traditional notion of competitive equilibrium in economics [16]. A multilateral equilibrium can be shown to exist under general conditions in our setting [14].

Moreover, the corresponding allocation is Pareto efficient, *i.e.*, there is no way to increase the utility of some peer without decreasing the utility of some other peer. A bilateral equilibrium, on the other hand, does not always exist, and, even when it exists, the allocation may not be (Pareto) efficient as the following examples illustrate.

Example 1. Consider a system with n peers and n files, for $n > 2$. Each peer i has file f_i and wants $f_{(i \bmod n)+1}$. With these utilities, no bilateral exchange can satisfy all peers, and a bilateral equilibrium does not exist.

Example 2. Consider a system with peers $\{1,2,3\}$ and files $\{f_1, f_2, f_3\}$. Peer i has file f_i and wants the other two files. The peer's utilities are, $V_1(d_{13}, d_{12}) = \ln(d_{13}) + 9\ln(d_{12})$, $V_2(d_{21}, d_{23}) = \ln(d_{21}) + 9\ln(d_{23})$, and $V_3(d_{32}, d_{31}) = \ln(d_{32}) + 9\ln(d_{31})$, where d_{ij} is the rate at which peer i downloads file f_j. The unique bilateral equilibrium is inefficient, because each peer is allocated a smaller rate of the file it values more.

These observations are intuitive: after all, exchange is far more restricted in a bilateral equilibrium than in a multilateral equilibrium. In the next section, we will focus on determining conditions under which a bilateral equilibrium yields a multilateral equilibrium.

We conclude this section by justifying our choice of pricing per peer instead of pricing per file. In the P2P setting we are considering, the two pricing schemes are equivalent in terms of equilibria, since the resource that is being priced is the upload capacity of a peer. A peer with multiple files will only upload his most "expensive" files at equilibrium. We chose to price per peer for simplicity. In the bilateral setting, pricing per file would require to have an exchange ratio $\gamma_{ij,fg}$ for each pair of files f, g that peers i, j can exchange, while pricing per peer only requires one exchange ratio for each pair of peers.

4   Robustness to Collusive Deviations

In this section we demonstrate that a bilateral equilibrium may not be robust to collusive deviations and show that a bilateral equilibrium corresponds to a multilateral equilibrium if and only if it is robust to deviations by coalitions of users. The following is a key step in establishing the relationship between bilateral and multilateral equilibrium. For clarity, all proofs are in the Appendix.

Proposition 1. *Consider a bilateral equilibrium with exchange ratios γ_{ij} for every pair of peers i, j. If there exist prices p_i for all $i \in N$ such that $\gamma_{ij} = p_i/p_j$ for all $i, j \in N$, then the bilateral equilibrium allocation is also a multilateral equilibrium allocation.*

This proposition is quite revealing: it shows that if exchange ratios are "fair," in the sense that they yield a unique price per peer, then the bilateral equilibrium allocation is also a multilateral equilibrium allocation, and thus is efficient.

We have already seen that a bilateral equilibrium need not exist and need not be Pareto efficient when it exists, whereas multilateral equilibria exist under general conditions and are Pareto efficient; thus, the two concepts are not equivalent. We now show that even an efficient bilateral equilibrium does not necessarily yield one price per peer, and thus is not always equivalent to a multilateral equilibrium.

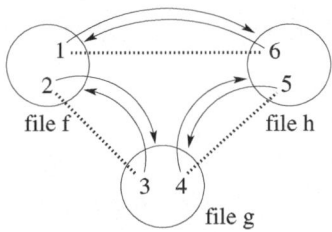

Fig. 2. Bilateral equilibrium for Ex.3. Peers $\{1,2\}$ have file f, peers $\{3,4\}$ have file g, and peers $\{5,6\}$ have file h. Solid arrows are drawn from a peer to its desired file, *e.g.*, 1 and 4 want h. Heavy dotted lines are between peers bartering at the unique bilateral equilibrium.

Example 3. There are 6 peers ($\{1,2,3,4,5,6\}$) and 3 files ($\{f,g,h\}$) in the system, with file allocation and demand as shown in Figure 2. The upload capacities of peers are $B_1 = 2$, $B_j = 1, \forall j \neq 1$. At the unique bilateral equilibrium, the following pairs exchange: $\{1,6\}$, $\{2,3\}$ and $\{4,5\}$, and the allocation is Pareto efficient. Thus for these pairs, the equilibrium exchange ratios must be $\gamma_{16} = 1/2$, $\gamma_{32} = 1$ and $\gamma_{54} = 1$.

The optimality conditions in the bilateral equilibrium must ensure that peer 1 does not wish to download from peer 5 instead of peer 6, which implies that we must have $\gamma_{15} \leq \gamma_{16} = 1/2$. Similarly, we must have $\gamma_{53} \leq \gamma_{54} = 1$, and $\gamma_{31} \leq \gamma_{32} = 1$. Note that we thus have $\gamma_{15}\gamma_{53}\gamma_{31} \leq 1/2$. However, this implies there do not exist prices per peer p_i such that $\gamma_{ij} = p_i/p_j$, since such a price vector would imply $\gamma_{15}\gamma_{53}\gamma_{31} = 1$. Thus the bilateral equilibrium cannot be a multilateral equilibrium.

Given the equilibrium exchange ratios of Example 3, peers $\{1,3,5\}$ can benefit by deviating together. By choosing upload rates $r'_{13} = 1/3, r'_{35} = 1/4, r'_{51} = 1/5$, while reducing upload rates to their original trading partners accordingly, each peer in $\{1,3,5\}$ obtains a download rate strictly larger than 1 (the download rate each of these peers gets at the bilateral equilibrium). For example, user 1 obtains a total download rate of $1/5$ (from user 5) plus $5/6$ (from user 6, who in turn gets $r'_{16} = 5/3$), which results in a rate greater than 1.

Inspired by this observation, we show next that if a bilateral equilibrium is robust to deviations by a coalition of peers, and if each peer is only uploading one file, then it corresponds to a multilateral equilibrium. We formalize this result adapting the notion of the *core* [16] to bilateral exchange. An allocation has the core property with respect to given exchange ratios if no coalition of peers can strictly improve the utility of all its members by bartering with peers outside the coalition, subject to the given exchange ratios. Inside the coalition, peers do not need to follow the exchange ratios, and they may allocate rates in any way subject to bandwidth constraints.

Definition 3. *Given exchange ratios $\gamma = (\gamma_{ij}, i, j \in N)$, an allocation r is feasible for a set of peers S with respect to γ if:*

(i) $r_{ijf} = 0$ *if* $f \notin F_i$;

(ii) $\sum_{j,f} r_{ijf} \leq B_i$ *for all* $i \in S$;

(iii) $\sum_{i \in S} \sum_f r_{jif} = \sum_{i \in S} \gamma_{ij} \sum_f r_{ijf}$, *for all* $j \notin S$.

The first condition ensures that all peers only upload files they have. The second condition ensures that peers in S do not exceed their upload constraints. The third ensures that exchanges between the coalition S and each peer outside S take place at the given exchange ratios.

Definition 4. *Given fixed exchange ratios* γ, *a coalition* S blocks *an allocation* r^* *with respect to* γ *if there exists a feasible allocation* r *for* S *with respect to* γ *such that* $V_i(\sum_j r_{jif}, f \in T_i) > V_i(\sum_j r^*_{jif}, f \in T_i)$ *for all* $i \in S$.

Definition 5. *The allocation* r *has* the core property *with respect to exchange ratios* γ *if it can not be blocked by any coalition of peers.*

We note that the usual definition of the core in microeconomics [16] does not allow exchange with agents outside the coalition. Our definition of the core is distinct and more appropriate to model collusion in a bilateral exchange setting, as it depends on the exchange ratios.

We first show that the core property is satisfied by any multilateral equilibrium. This is a standard result from microeconomic theory [16]. However, our result is more general, since our core definition allows a coalition to exchange with peers outside the coalition, and thus there are more feasible allocations which may potentially block the multilateral equilibrium allocation.

Proposition 2. *Any multilateral equilibrium allocation has the core property with respect to the equilibrium exchange ratios* $\gamma_{ij} = p_i/p_j$.

We conclude that a bilateral equilibrium that does not have the core property can not correspond to a multilateral equilibrium. We next show that, when each peer is uploading one file, a bilateral equilibrium with the core property is a multilateral equilibrium. The insight is similar to Example 3: it can be shown that if no price vector exists such that $\gamma_{ij} = p_i/p_j$, then there must exist users $i_1, i_2, ..., i_k$ such that $\prod_{i=1}^{k} \gamma_{i,(i \bmod k)+1} < 1$. In that case, there is a coalition of k peers that can block the allocation.

Proposition 3. *Suppose* $|F_i| = 1$ *for all* $i \in N$. *If a bilateral equilibrium allocation* r^* *with exchange ratios* γ *such that* $\sum_j r^*_{jif} > 0$ *for all* $i \in N, f \in T_i$ *has the core property, then it is also a multilateral equilibrium allocation.*

A corollary of Proposition 3 is that if a bilateral equilibrium has the core property, then its allocation is Pareto efficient. Proposition 3 requires that $|F_i| = 1$ for all $i \in N$. The result holds more generally if each peer is uploading a unique file at the bilateral equilibrium (peers may have more files). Whether it holds for the general case where peers upload multiple files at the bilateral equilibrium remains an open problem.

5   Conclusions

This paper analyzes the efficiency and robustness gains that are enabled by price-based multilateral exchange. We identify the condition that a bilateral equilibrium needs to satisfy in order to correspond to a multilateral equilibrium. These results help clarify the tradeoffs inherent in choosing between bilateral and multilateral exchanges: simplicity in the former, and efficiency and robustness gains in the latter.

Our novel theoretical results provide insight into the gap between bilateral and multilateral exchange. Even though the two exchanges are compared theoretically in terms of equilibria in this paper, it is possible to design a system that practically realizes the benefits of multilateral exchange. In particular, since it is hard to know the equilibrium prices in advance, peers can update their prices according to supply and demand. A system for currency-backed content exchange is presented in [15].

References

[1] Bangeman, E.: P2P responsible for as much as 90 percent of all 'Net traffic. ArsTechnica (September 3, 2007)

[2] Adar, E., Huberman, B.: Free riding on Gnutella. First Monday 5(10) (2000)

[3] Hughes, D., Coulson, G., Walkerdine, J.: Free riding on Gnutella revisited: The bell tolls? IEEE Distributed Systems Online 6(6) (2005)

[4] Cohen, B.: Incentives build robustness in BitTorrent. In: Proc. Workshop on Economics of Peer-to-Peer Systems (2003)

[5] Piatek, M., Isdal, T., Anderson, T., Krishnamurthy, A., Venkataramani, A.: Do incentives build robustness in BitTorrent? In: Proc. Symposium on Networked Systems Design and Implementation (2007)

[6] Wu, F., Zhang, L.: Proportional response dynamics leads to market equilibrium. In: Proc. Symposium on Theory of Computing (2007)

[7] Jun, S., Ahamad, M.: Incentives in BitTorrent induce free riding. In: Proc. Workshop on the Economics of Information Security (2005)

[8] Locher, T., Moor, P., Schmid, S., Wattenhofer, R.: Free riding in BitTorrent is cheap. In: Proc. HotNets (2006)

[9] Sirivianos, M., Park, J.H., Chen, R., Yang, X.: Free-riding in BitTorrent networks with the large view exploit. In: Proc. International Workshop on Peer-to-Peer Systems (2007)

[10] Guo, L., Chen, S., Xiao, Z., Tan, E., Ding, X., Zhang, X.: Measurements, analysis, and modeling of BitTorrent-like systems. In: Proc. Internet Measurement Conference (2005)

[11] Gupta, M., Judge, P., Ammar, M.: A reputation system for peer-to-peer networks. In: Proc. Workshop on Network and Operating System Support for Digital Audio and Video (2003)

[12] Vishnumurthy, V., Chandrakumar, S., Sirer, E.G.: KARMA: A secure economic framework for P2P resource sharing. In: Proc. Workshop on the Economics of Information Security (2003)

[13] Sirivianos, M., Park, J.H., Yang, X., Jarecki, S.: Dandelion: Cooperative content distribution with robust incentives. In: Proc. USENIX Technical Conference (2007)

[14] Aperjis, C., Johari, R.: A peer-to-peer system as an exchange economy. In: Proc. GameNets (2006)

[15] Aperjis, C., Freedman, M.J., Johari, R.: The role of prices in peer-assisted content distribution. Technical report

[16] Mas-Colell, A., Whinston, M.D., Green, J.R.: Microeconomic Theory. Oxford University Press, Oxford (1995)

A Proofs

Proof of Proposition 1: Substituting $\gamma_{ij} = p_i/p_j$ in the fourth set of constraints of the Bilateral Peer Optimization problem, we get the constraints $p_j \sum_f r_{jif} = p_i \sum_f r_{ijf}$ for all j. The Multilateral Peer Optimization problem has a larger feasible region; however, the two optimization problem have the same optimal value. In particular, if r is feasible for the Multilateral Optimization problem of peer i, then we can construct $\bar{r}$ such that (i) $\bar{r}$ is feasible for the Bilateral Optimization problem of peer i, and (ii) $\bar{r}$ gives the same utility as r to user i. For example, we can achieve this by setting $\bar{r}_{jif} = r_{jif}$ for all j, f and then choosing $\bar{r}_{ijf}$ for each j such that $\sum_f \bar{r}_{ijf} = (p_j/p_i) \sum_f \bar{r}_{jif}$. This shows that an optimal solution for the Bilateral Peer Optimization problem is also optimal for the Multilateral Peer Optimization problem, and concludes the proof. ∎

Proof of Proposition 2: Consider a competitive equilibrium and suppose that there exists a coalition S that blocks it and let r be the corresponding rate allocation. Then by Definition 4 and the Multilateral Optimization problem, $\sum_{j,f} p_j \cdot r_{jif} > p_i \cdot B_i, \forall i \in S$. Summing over all $i \in S$, $\sum_{i \in S} \sum_{j \in S} p_j \cdot \sum_f r_{jif} + \sum_{i \in S} \sum_{j \notin S} p_j \cdot \sum_f r_{jif} > \sum_{i \in S} p_i \cdot B_i$. Since $\gamma_{ij} = p_i / p_j$, condition (iii) of Definition 3 becomes $p_j \cdot \sum_{i \in S} \sum_f r_{jif} = \sum_{i \in S} p_i \cdot \sum_f r_{ijf}$ for all $i \in S$, $j \notin S$. Combining the two conditions and rearranging, $\sum_{i \in S} p_i \sum_{j,f} r_{ijf} > \sum_{i \in S} p_i \cdot B_i$, which contradicts condition (ii) of Definition 3. ∎

Proof of Proposition 3: We first show that the exchange ratios in a bilateral equilibrium only depend on the files being exchanged, not on the peers' identities. Suppose that $f \in F_{i_1}$, $f \in F_{i_2}$, $g \in F_{j_1}$ and $g \in F_{j_2}$ and at the bilateral equilibrium, peers i_1 and j_1 exchange f and g, and i_2 and j_2 exchange f and g. Then,

$$\gamma_{i_1,j_1} \geq \gamma_{i_1,j_2} \; ; \; \gamma_{j_1,i_1} \geq \gamma_{j_1,i_2} \; ; \; \gamma_{i_2,j_2} \geq \gamma_{i_2,j_1} \; ; \; \gamma_{j_2,i_2} \geq \gamma_{j_2,i_1},$$

since i_1 exchanges with j_1 at equilibrium, not j_2; j_1 exchanges with i_1, not i_2; etc. Combining these inequalities with the fact $\gamma_{ij} = 1/\gamma_{ji}$, we conclude that $\gamma_{i_1,j_1} = \gamma_{i_1,j_2} = \gamma_{i_2,j_1} = \gamma_{i_2,j_2}$. When there is exchange between files f and g in a bilateral equilibrium we define γ_{fg} to be the unique value of the exchange ratio between files f and g.

We define the *bilateral equilibrium exchange graph* to have a node for every file, and an edge between two files if those files are exchanged at the bilateral equilibrium. Suppose that for every cycle $f_1, f_2, ..., f_k, f_1$ in the bilateral equilibrium exchange graph, we have $\prod_{i=1}^{k} \gamma_{f_i, f_{i+1}} = 1$.[1] Then we get a unique price for each file in the following way. We start from a file (say f_1) whose price we set equal to 1 and then take a minimum spanning tree of the graph. We move along the edges of this tree and set prices for other files, so that the exchange ratios are satisfied. In this way we get prices p_f for each file f. Since the optimization problem of peer i with $f \in F_i$ is not affected by γ_{fg} if $g \notin T_i$, we can derive prices per peer by setting $p_i = p_f$ for $f \in F_i$ which do not change the optimal rate allocations for peers. Thus, by Proposition 1, the bilateral equilibrium allocation is also a multilateral equilibrium allocation.

To complete the proof, it suffices to show that if there is a cycle $f_1, f_2, ..., f_k, f_1$ in the bilateral equilibrium exchange graph with $\prod_{i=1}^{k} \gamma_{f_i, f_{i+1}} < 1$, then there is a coalition that blocks the bilateral equilibrium allocation. Let i be some peer that uploads file f_i, and downloads file f_{i+1} at the bilateral equilibrium. It can be shown that the coalition $S = \{1, ..., k\}$ blocks the bilateral equilibrium allocation by demonstrating that there is a way to increase the rates $r_{i,i-1,f_i}$ for $i = 1, ..., k$ so that the utilities of all peers in S strictly increase. Due to space limitations, here we only sketch the proof for the case that $f_{i-1} \notin T_i$ for all $i \in S$. We provide the complete proof in [15].

If $f_{i-1} \notin T_i$ for all $i \in S$, then at the bilateral equilibrium $r_{i,i-1,f_i}^* = 0$ for all $i \in S$. By sending $r_{i,i-1,f_i}$ to peer $i-1$, peer i reduces the rate he gets from outside S by $\gamma_{f_i,f_{i+1}} \cdot r_{i,i-1,f_i}$. So, the coalition increases i's utility if and only if the rate he receives from S, *i.e.*, $r_{i+1,i,f_{i+1}}$, is greater than $\gamma_{f_i,f_{i+1}} \cdot r_{i,i-1,f_i}$. To show that S blocks r^*, it suffices to find $r_{i+1,i,f_{i+1}} \leq B_{i+1}$, such that $r_{i+1,i,f_{i+1}} > \gamma_{f_i,f_{i+1}} \cdot r_{i,i-1,f_i}$, for all $i \in S$. This is possible because $\prod_i \gamma_{f_i,f_{i+1}} < 1$. In particular, we can choose small $\delta, \varepsilon > 0$, and set: $r_{1,k,f_1} = \delta$, $r_{i+1,i,f_{i+1}} = \gamma_{f_i,f_{i+1}} r_{i,i-1,f_i} + \varepsilon$, for all $i \in S$. ∎

---

[1] We denote by f_{i+1} and f_{i-1} the files after and before file f_i with respect to the cycle $f_1, f_2, ..., f_k, f_1$.

Auctions for Resource Allocation in Overlay Networks

Pablo Belzarena[1], Andrés Ferragut[1,2], and Fernando Paganini[2]

[1] Universidad de la República
[2] Universidad ORT
Montevideo, Uruguay

Abstract. The paper studies the problem of allocating bandwidth resources of a Service Overlay Network, to optimize revenue. Clients bid for network capacity in periodically held auctions, under the condition that resources allocated in an auction are reserved for the entire duration of the connection, not subject to future contention. This makes the optimal allocation coupled over time, which we formulate as a Markov Decision Process (MDP). Studying first the single resource case, we develop a receding horizon approximation to the optimal MDP policy, using current revenue and the expected revenue in the next step to make bandwidth assignments. A second approximation is then found, suitable for generalization to the network case, where bids for different routes compete for shared resources. In that case we develop a distributed implementation of the auction, and demonstrate its performance through simulations.

1   Introduction

In recent years, many types of "overlay" networks have been proposed for the Internet. These overlays include content delivery networks, peer-to-peer file sharing, some voice-over-IP services, and testbed networks such as PlanetLab. A particular architecture called Service Overlay Network (SON) [6] has been proposed to deploy value-added Internet services with end-to-end quality of service (QoS). The basic components of the SON architecture are service gateways located in domain boundaries, and a network of tunnels acquired from the underlying domains with guaranteed bandwidth. Through this overlay, a client with good local connectivity in one domain can secure a high QoS connection with a remote domain. The overlay operator invests in the infrastructure and leased bandwidth to offer high-value services, for instance through distributed content servers; a profitable sale of this capacity is thus essential. In this paper we study an access control policy based on auctions for this purpose, where users bid for a service connection and the network gives access to the best bids to maximize its payoff, subject to the QoS constraints for the admitted clients.

Resource sharing policies based on auctions have been analyzed by different previous works [7],[5],[8],[10],[3]. One main difference in our approach is that we impose the condition that once bandwidth has been allocated in an auction, the successful bidder is guaranteed to hold it for the duration of his/her connection. Previous work on auctions allows future bidders to compete with incumbent

E. Altman and A. Chaintreau (Eds.): NET-COOP 2008, LNCS 5425, pp. 9–16, 2009.

ones, albeit given the latter some advantage. Our application scenario does not allow this: consider selling video-on-demand content about 100 minutes long, in auctions every 5 minutes. A consumer will not purchase the service if he/she faces the risk of losing the connection close to the end of the movie.

Reserving bandwidth over multiple auctions means that the operator must assume the risk of future bids. Optimizing revenue with this risk becomes a stochastic dynamic optimization problem, that we formulate as a Markov decision process (MDP) [1,9]. The optimization involves a tradeoff between the revenue of the current auction, and the expected value of bids the operator will miss in future auctions if it runs out of resources. In Section 2 we analyze the above problem in the case of a single link, and develop a series of approximations to the optimal policy. The aim of our approximations is to allow a distributed implementation of the policy over an arbitrary network topology, where bids are received at the edge for end-to-end services, and the network optimizes the overall revenue. This is described in Section 3, and the method is evaluated by simulation. Conclusions given in Section 4.

2   Auctions for One Link

In this section we consider auctions for one access link. We make a few simplifying assumptions: all consumers bid for the same amount (unit) of bandwidth, and the link has capacity for C such connections. Bids are collected for a period of time T, and an auction is held; we assume for simplicity that the number N of bids is given, and bids are drawn from a known probability distribution. We denote by $b_k^{(1)} \geq b_k^{(2)} \geq \cdots \geq b_k^{(N)}$ the ordered bids in decreasing order that participate in the auction at time kT. The bids are for the entire duration of the service, and this is a first-price auction: the admitted users will pay their bid.

Let a_k represent the number of admitted connections in auction k. The revenue from this auction is then

$$U_{b_k}(a_k) := \sum_{i=1}^{a_k} b_k^{(i)}. \tag{1}$$

The function $U_b(a)$ is defined above for integer values of a; it is also convenient to extend it to a function of $a \in \mathbb{R}$, by linear interpolation. The latter function is increasing and concave in a. We also define $\overline{U}(a) = E[U_b(a)]$, where the expectation is over the distribution of the bids b; i.e., we replace the current bids in (1) by their expectation. This is also increasing, piecewise linear and concave.

We will model the connection durations as independent exponential random variables, of mean $1/\mu$. Therefore at the end of the period T each connection has probability $p := e^{-\mu T}$ of remaining active for the following period. While in practice one would have more deterministic information of the service duration, the above allows for a Markovian treatment of the allocation problem.

Let x_k denote the number of connections active at $t = kT^-$, i.e. before the k-th auction. The system admits a_k new connections, $0 \leq a_k \leq C - x_k$, taking the total to $x_k + a_k$. By the next auction period, $t = (k+1)T^-$, the number of active connections x_{k+1} follows then a binomial distribution with parameters

$x_k + a_k$ and p. Specifically, $P[x_{k+1} = i | x_k, a_k] = \binom{x_k + a_k}{i} p^i (1-p)^{x_k + a_k - i}$. We are ready to state our design objective.

Optimal revenue problem: Maximize $\lim_n \frac{1}{n} \sum_{k=0}^{n-1} E[U_{b_k}(a_k)]$.

Here the expectation is over two sources of randomness: the vector of bids b_k and the departure process. The constraints are $0 \leq a_k \leq C - x_k$ where x_k follows the binomial transition dynamics defined above. We can also consider the discounted version: Maximize $\sum_{k=0}^{\infty} \rho^k E[U_{b_k}(a_k)]$, where $0 < \rho < 1$.

Both of these are Markov Decision Processes (MDPs) [2,9]. The *state* at time k is $s_k = (x_k, b_k)$, i.e. the current occupation and the incoming bids. Based on this state, the *action* $a_k = a(s_k)$ decides how many bids to accept. Solving the MDP requires finding the *policy* $a(s)$ that results in a minimum cost. In the discounted case $\rho < 1$, this policy satisfies the Bellman equation

$$V^*(x_0, b) = \max_{a \in \mathcal{A}_s} \{U_b(a) + \rho E[V^*(x_1, b')]\}, \tag{2}$$

where V^* is the value function and the expectation is taken over the binomial distribution of $x_1 | (x_0, a)$ and the distribution of the next bid b'. The state-dependent constraints are $\mathcal{A}_s = \{0 \leq a \leq C - x_0\}$. For $\rho = 1$, V^* satisfying (2) is no longer the optimal cost, but (2) still characterizes the optimal action $a(s)$.

It is in general difficult to solve the Bellman equation; a commonly used strategy is the *value iteration* $V_{m+1}(x_0, b) := \max_{a \in \mathcal{A}_s} \{U_b(a) + \rho E[V_m(x_1, b')]\}$; starting with an arbitrary $V_0(s)$, $V_m(s)$ converges to $V^*(s)$, and the corresponding maximizing action converges to the optimal action [2].

Receding horizon approximation. We will use initial steps of the value iteration to approximate the optimal policy. Starting from $V_0 \equiv 0$, we have

$$V_1(x_0, b) = \max_{a \leq C - x_0} U_b(a) = U_b(C - x_0). \tag{3}$$

This first step gives the "myopic" policy $a = C - x_0$, that sells all available capacity disregarding the future. To improve on it, we take a second step in the value iteration:

$$V_2(x_0, b) = \max_{a \leq C - x_0} \{U_b(a) + \rho E[V_1(x_1, b')]\}$$

$$= \max_{a \leq C - x_0} \{U_b(a) + \rho E[U_{b'}(C - x_1)]\}$$

$$= \max_{a \leq C - x_0} \{U_b(a) + \rho E_{x_1} \overline{U}(C - x_1)]\}. \tag{4}$$

In (4), the expectation over the random bid b' is included in $\overline{U}$ defined above; the remaining expectation is over $x_1 \sim \text{Bin}(x_0 + a, p)$. The policy (4) can be given a *receding horizon* interpretation: the decision considers the current revenue plus the expected revenue of looking one step ahead, assuming all available capacity will be sold off at that time. This decision is then applied recursively; thus the future is taken into account at a limited level of complexity.

The first term in (4) increases with a. To characterize the second, we rewrite it as follows. Consider the function $W(i) = \overline{U}(C) - \overline{U}(C - i)$, piecewise linear and

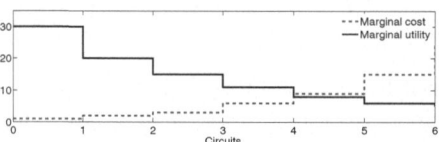

Fig. 1. Marginal utility versus marginal cost

convex in i. Indeed, the increments $W(i+1) - W(i) = E[b^{(C-i)}]$, (expectation of the $(C-i)$-th largest offer) are increasing in i. Now define $\overline{W}(x) = E[W(x_1)]$, where $x_1 \sim \text{Bin}(x, p)$, again linearly interpolated for non-integer x. The next statement follows from a stochastic comparison argument, omitted for brevity.

Proposition 1. $\overline{W}(x)$ *is increasing and convex in* x.

With this notation, our one-step ahead optimization can be rewritten as

$$\max_{a \leq C - x_0} U_b(a) - \rho \overline{W}(x_0 + a), \tag{5}$$

a convex optimization problem. In it, $\overline{W}(x_0 + a)$ plays the role of a *cost function* that makes the decision at the current time "internalize" the impact on future decisions. $\overline{W}(x_0 + a)$ measures the expected loss of revenue in the auction at time T from having left $x_0 + a$ occupied circuits at time $t = 0^+$.

It is now straightforward to determine the optimal policy $a(x_0, b)$. To optimize over a we plot the derivatives of $U_b(a)$ and $\rho \overline{W}(x_0 + a)$ (marginal utilities and costs) and look for a crossing point. This is depicted in Figure 1. The marginal utilities are just the current bids in decreasing order. The marginal costs represent the value of leaving one more free circuit for the next auction.

The increasing marginal costs $\overline{w}_i = \rho(\overline{W}(i) - \overline{W}(i-1))$ act as successive thresholds for accepting bids. The acceptance policy is the value a such that for $i = x_0 + a$ we have $b^{(1)} \geq b^{(2)} \geq \cdots \geq b^{(a)} \geq \overline{w}_i > b^{(a+1)}$. In words: to accept a bids, the *lowest* one must exceed $\overline{w}_i$ for $i = x_0 + a$. To accept one more, we require a *more demanding* threshold $\overline{w}_{i+1}$ on this (smaller) bid[1].

From the binomial distributions $j_1 \sim \text{Bin}(i, p)$, $j_2 \sim \text{Bin}(i-1, p)$ and some combinatorics we obtain for the thresholds the expression

$$\overline{w}_i = \rho[E(W(j_1)) - E(W(j_2))] = \rho \, p \sum_{l=0}^{i-1} E(b^{(C-l)}) \binom{i-1}{l} p^l (1-p)^{i-1-l}. \tag{6}$$

Based on knowledge of ρ, p, and the distribution of bids, this expression could be calculated offline and used for carrying out auctions with the policy (4).

Examples: We evaluate the previous results in a few simple cases (for $\rho = 1$).

For $C = 1$, there is a single link cost $\overline{w}_1 = pE(b^{(1)})$, that acts as an admission threshold for bids received when the circuit becomes empty. For instance in the case of N bids, uniformly distributed in $[0, b_{max}]$, we have $\overline{w}_1 = p\frac{N}{N+1} b_{max}$.

---

[1] Since b's are random, the curves of Fig. 1 will generically cross at a single point.

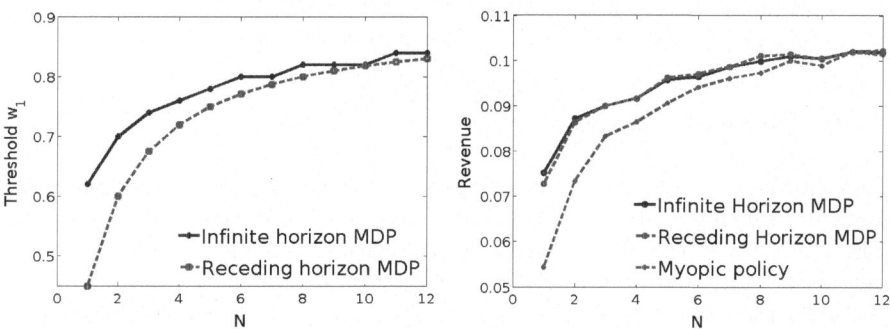

Fig. 2. Optimal MDP, receding horizon and myopic policies. $C = 1$, $p = 0.1$.

If the link has capacity $C = 2$, there are two marginal costs: $\overline{w}_1 = pE(b^{(2)})$ for occupying the first connection, and $\overline{w}_2 = p(E(b^{(2)})(1 - p) + E(b^{(1)})p)$ for occupying the second. Again, for uniform in $[0, b_{max}]$ bids this becomes

$$\overline{w}_1 = p\,\frac{N - 1}{N + 1}\,b_{max}; \qquad \overline{w}_2 = p\,\frac{(N - 1)(1 - p) + Np}{N + 1}\,b_{max}.$$

We now compare by simulations our receding horizon policy with the optimal infinite-horizon MDP, in the case of one circuit ($C = 1$). In this simple case, the latter is also a threshold policy on the bids, but the optimal threshold does not have a simple formula; we computed it numerically through the value iteration algorithm from [4]. On the left in Figure 2 we show the acceptance thresholds for both policies: we see the infinite horizon threshold is more demanding. On the right we show the average utility obtained empirically by simulation of these two policies. Results are very similar. Therefore, in this case we have managed to extract almost the optimal utility just be looking one-step ahead with the policy. On the other hand, if we apply the myopic policy that always fills the link, the right-hand side plot shows there is a clear loss in utility.

A further approximation. Our ultimate goal is to perform an auction over a general network, where bids appear at different routes and are coupled by scarce resources at links. The stochastic calculations involved in (6) appear difficult to generalize, so we adopt a second approximation, replacing the function $\overline{W}(x)$ in (5) by something easier to compute. Namely, define $\phi(x) = W(E[x_1])$ for $x_1 \sim \text{Bin}(x, p)$. Since $W(\cdot)$ is convex, this underestimates the one-step cost from before, $\phi(x) \leq \overline{W}(x)$. Nevertheless, if C is large the binomial distribution will be concentrated around its mean and the error is moderate. In return, we have the simple expression $\phi(x) = W(px) = \overline{U}(C) - \overline{U}(C - px)$. This is still piecewise linear and convex, but easier to compute. The second approximation to the optimal policy is given by $\max_{a \leq C - x_0} U_b(a) - \rho\phi(x_0 + a)$.

Equivalently, we can rewrite the above as the convex program

$$\begin{aligned} \max\quad & U_b(a) + \rho\overline{U}(z), \\ \text{subject to}\quad & x_0 + a \leq C, \quad p(x_0 + a) + z \leq C. \end{aligned} \tag{7}$$

At the optimum, the slack variable z will satisfy the constraint with equality, i.e. $z = C - p(x_0 + a)$ is the expected allocation in the next interval.

3   The Network Case

In the following, l indexes the network links, r the routes across the network. R denotes the routing matrix, $R_{lr} = 1$ iff route r includes link l. $c = (c_l)$ is the vector of link capacities. We describe the allocation decision at time $t = 0$.

Define column vectors x_0, a, and z, whose coordinates per route r denote respectively the rate x_0^r from previous occupation, the rate allocation a^r at the current auction, and the expected allocation z^r in the following auction ($t = T$). We also define the piecewise linear utility $U_{b^r}(a^r)$ based on current bids (1), and the utility $\overline{U}_r(z^r)$ based on expected bids. Let $p^r = e^{-\mu^r T}$ be the probability that a connection active at $t = 0$ will remain active at $t = T$; $P = \mathrm{diag}(p^r)$ is the corresponding diagonal matrix. Thus, $P(a + x_0)$ is the expected input rate vector at $t = T^-$. The network generalization of (7) is

$$\max \sum_r \left[U_{b^r}(a^r) + \rho \overline{U}_r(z^r) \right],$$

$$\text{subject to} \quad R(a + x_0) \le c; \qquad RP(a + x_0) + Rz \le c. \tag{8}$$

This optimization can also be rewritten as one for the current allocation a with a convex cost function $\phi(x_0 + a)$ that represents the optimization in z. However, here the cost function would be coupled over the network. A better way to solve (8) is by duality. Consider the Lagrangian $L(a, z, \alpha, \beta)$ given by

$$L = \sum_r \left[U_{b^r}(a^r) + \rho \overline{U}_r(z^r) \right] + \alpha^T(c - R(a + x_0)) + \beta^T(c - Rz - RP(a + x_0))$$

$$= \sum_r [U_{b^r}(a^r) - (q^r + p^r v^r)a^r] + [\rho \overline{U}_r(z^r) - v^r z^r] + \alpha^T(c - Rx_0) + \beta^T(c - RPx_0).$$

Here, α and β are the vectors of Lagrange multipliers (prices) for each of the two constraints, and we have defined the aggregate prices per route $q = R^T \alpha$, $v = R^T \beta$. We can solve the convex program through a dual, gradient projection algorithm similar to those used in the congestion control literature [11], but with additional prices. The algorithm takes the following form (in continuous time):

$$a^r = \arg\max_{a^r}[U_{b^r}(a^r) - (q^r + p^r v^r)a^r]; \qquad \dot{\alpha} = [R(a + x_0) - c]_\alpha^+;$$

$$z^r = \arg\max_{z^r}[\rho \overline{U}_r(z^r) - v^r z^r]; \qquad \dot{\beta} = [RP(a + x_0) + Rz - c]_\beta^+.$$

An important difference with the congestion control case is that here the algorithm should take place in the *control plane* prior to the allocation of resources. Despite this difference, we can still obtain a *distributed* computation by message passing between the network links, and a broker entity at the ingress node of route r. The maximization for a^r amounts simply to comparing the bids with

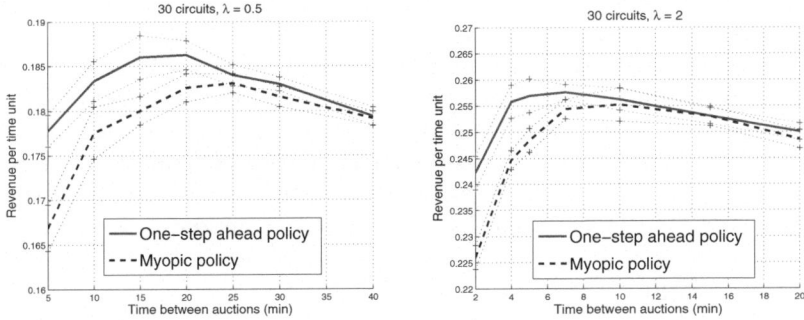

Fig. 3. One link situation: 30 circuits, offer arrival rate $\lambda = 0.5$ (left) and $\lambda = 2$ (right). Dotted lines are 5% confidence intervals.

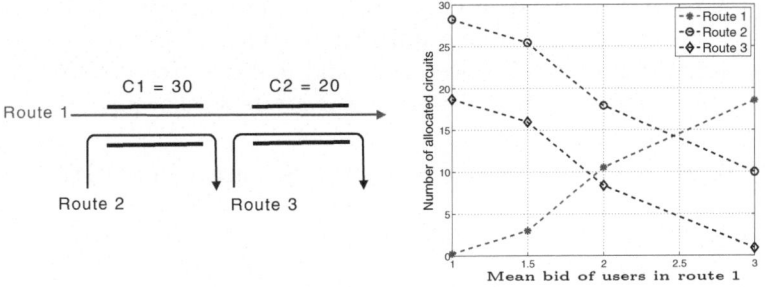

Fig. 4. Linear network, allocation as a function of the mean bid $\bar{b}$ of route 1

the threshold price $q^r + p^r v^r$. Solving for z^r involves the expected bids, and the price v^r/ρ. We must allow time after closing the auction for this distributed algorithm to settle on an allocation vector a.

It is useful to compare this algorithm with an implementation of the myopic policy which optimizes the first term in (8); the latter would involve only the variables a, α, but a very similar overhead in terms of message passing.

Simulations. To evaluate the proposed algorithm we present some simulation studies, carried out on a flow-level simulator we developed in Java.

First, we compare the one-step ahead policy (7) with the myopic one (3) in a single link case, with $C = 30$ circuits. Auctions take place every T minutes, and bids arrive periodically with fixed intensity λ, totalling $N = \lambda T$ bids per auction. Bids are independent and uniformly distributed in $[0, 1]$; rejected bids are discarded. Job duration is $exp(\mu)$ with $1/\mu = 100$ minutes.

We simulate both policies and compare the respective revenues, for several values of T which is a critical design parameter: enlarging T will allow more bids per auction, hence better ones, but decrease the auction rate. Results are shown in Fig. 3 for two different values of λ. In both cases, the one-step ahead policy outperforms the myopic one, as expected; and as remarked before, the gain is achieved with minor extra overhead.

Our second study uses the linear network topology of Fig. 4. In this case, users in the long route 1 should pay more to be allocated a circuit, since it occupies two links. We used $N = 10$ uniform bids with mean 1 for the short routes, and with mean $\bar{b}$ for route 1. Fig. 4 shows the mean number of connections admitted by the auction algorithm, for several values of $\bar{b}$. If $\bar{b} = 1$, the users in route 1 starve, but for $\bar{b} \approx 2$ all routes receive a comparable fraction of resources.

4   Conclusions

In this work we proposed an auction mechanism to assign resources in overlay networks. We formulated the problem of maximizing operator revenue, and found near-optimal policies that can be computed via convex optimization, and allow a distributed implementation over a network. In future work we will study some natural extensions: random numbers of bids per auction, bidding for heterogeneous services which consume different amounts of bandwidth, learning an unknown distribution of bids over time, and multiple-step extensions to the receding horizon policy. We will also study the strategic aspects of the auction.

Acknowledgment. This work was supported by a grant from PDT-Uruguay.

References

1. Altman, E.: Applications of Markov Decision Processes in Communication Networks: a Survey. In: Feinberg, E., Shwartz, A. (eds.) Markov Decision Processes, Models, Methods, Directions, and Open Problems, pp. 488–536. Kluwer, Dordrecht (2001)
2. Bertsekas, D.P.: Dynamic Programming: Deterministic and Stochastic Models. Prentice-Hall, Englewood Cliffs (1987)
3. Dramitinos, M., Stamoulis, G.D., Courcoubetis, C.: An auction mechanism for allocating the bandwidth of networks to their users. Computer Networks 51 (2007)
4. Chads, I., Cros, M.-J., Garcia, F., Sabbadin, R.: Markov Decision Process Toolbox for MATLAB,
   `http://www.inra.fr/internet/Departements/MIA/T//MDPtoolbox/`
5. Dramitinos, M., Stamoulis, G., Courcoubetis, C.: Auction-based resource reservation in 2.5/3G Networks. Mobile Networks and Apps. 9, 557–566 (2004)
6. Duan, Z., Zhang, Z.-L., Hou, Y.T.: Service Overlay Networks: SLAs, QoS and Bandwidth Provisioning. IEEE/ACM Trans. on Networking, 870–883 (2003)
7. Lazar, A.A., Semret, N.: Design and Analysis of the Progressive Second Price auction for Network Bandwidth Sharing. Telecommunications Systems (2000)
8. Maillé, P., Tuffin, B.: Pricing the internet with multibid auctions. IEEE/ACM Trans. Netw. 14(5), 992–1004 (2006)
9. Puterman, M.L.: Markov Decision Processes. Wiley, New Jersey (2005)
10. Reichl, P., Wrzaczek, S.: Equilibrium Market Prices for Multi-Period Auctions of Internet Resources. In: MMB 2004, pp. 25–34 (2004)
11. Srikant, R.: The Mathematics of Internet Congestion Control. Birkhäuser, Basel (2004)

Marketing in a Random Network

Hamed Amini[1], Moez Draief[2], and Marc Lelarge[1]

[1] INRIA & ENS, Paris
[2] Imperial College, London

Abstract. Viral marketing takes advantage of preexisting social networks among customers to achieve large changes in behaviour. Models of influence spread have been studied in a number of domains, including the effect of "word of mouth" in the promotion of new products or the diffusion of technologies. A social network can be represented by a graph where the nodes are individuals and the edges indicate a form of social relationship. The flow of influence through this network can be thought of as an increasing process of active nodes: as individuals become aware of new technologies, they have the potential to pass them on to their neighbours. The goal of marketing is to trigger a large cascade of adoptions. In this paper, we develop a mathematical model that allows to analyze the dynamics of the cascading sequence of nodes switching to the new technology. To this end we describe a continuous-time and a discrete-time models and analyse the proportion of nodes that adopt the new technology over time.

Keywords: Models of contagion, random graphs.

1 Introduction

With consumers showing increasing resistance to traditional forms of advertising, marketers have turned to alternate strategies like *viral marketing*. Viral marketing exploits existing social networks by encouraging customers to share product information with their friends. Social networks are graphs in which nodes represent individuals and edges represent relations between them. To illustrate viral marketing, consider a company that wishes to promote its new instant messenger (IM) system [10]. A promising way would be through popular social network such as Myspace: by convincing several persons to adopt the new IM system, the company can obtain an effective marketing campaign and diffuse the new system over the network.

If we assume that "convincing" a person to "spread" the new technology costs money, then a natural problem is to detect the influential members of the network who can trigger a cascade of influence in the most effective way [2], [7]. In this work, we consider a slightly different problem: the marketer has no knowledge of the social network. Hence he will not be able to detect the most influential individuals and his only solution is to "convince" a fraction of the total population. However, the marketer can still use the structure of the underlying network by targeting the neighbours of the adopters. There are a number of

E. Altman and A. Chaintreau (Eds.): NET-COOP 2008, LNCS 5425, pp. 17–25, 2009.
© Springer-Verlag Berlin Heidelberg 2009

incentive programs around this idea: each time an individual chooses the new technology, he is given the opportunity to send e-mail to friends with a special offer; if the friend goes on to buy it, each of the individuals receives a small cash bonus.

In this paper, we develop a mathematical model that allows to analyze the dynamics of the cascading sequence of nodes switching to the new technology. To this end we describe a continuous-time and a discrete-time models and analyze the proportion of nodes that adopt the new technology over time. In the continuous setting we derive a general bound for the proportion of new adopters in terms global graph properties, namely the spectral radius and the minimum degree. In the discrete setting we show that the proportion of new adapters is the solution of a fixed point equation. To this end we examine the case of regular trees, and prove that our approach carries over to random regular graphs. We extend our model to the general threshold model [7] and to sparse random graphs. We conclude by presenting a framework that enables the control of the marketing policy and discuss other possible applications.

2   Model

We consider a set of n agents represented by an undirected graph structure $G = (V, E)$ accounting for their interaction. For $i, j \in V$, we write $i \sim j$ if $(i, j) \in E$ and we say that agents i and j are neighbours. As in [14], we consider binary models where each agent may choose between two possible strategies that we denote by A and B. Let us introduce a game-theoretic diffusion model proposed by Morris [12]: Whenever two neighbours in the graph opt for strategy A they receive a payoff q_A, if they both choose B they receive a payoff q_B, and they receive nothing if they choose opposite strategies. The payoff of an agent corresponds to the sum of its payoffs with each of its neighbours.

Initially all nodes play A except for a small number of nodes that are forced to adopt strategy B. The nodes that started with strategy A will subsequently apply best-response updates. More precisely, these nodes will be repeatedly applying the following rule: switch to B if enough of your neighbours have already adopted B. There can be a cascading sequence of nodes switching to B such that a network-wide equilibrium is reached in the limit. This equilibrium may involve uniformity with all nodes adopting B or it may involve coexistence, with the nodes partitioned into a set adopting B and a set sticking to A.

The state of agent i is represented by X_i; $X_i = 0$ if player i plays strategy A and $X_i = 1$ otherwise. Hence $\sum_{j \sim i} X_j$ is the number of neighbours of i playing strategy B and $\sum_{j \sim i}(1 - X_j)$ is the number of neighbors of i playing strategy A.

We now describe the economic model for the agents. Recall that the payoff for a $A - A$ edge is q_A, for a $B - B$ edge is q_B and for a $A - B$ edge is 0. We assume that if an agent chooses A, his payoff is just the sum of the payoffs obtained on each of his incident edges but if he chooses B, his payoff is the sum of these payoffs increased by an amount $u \geq 0$ plus a bonus of $r \geq 0$. Now the total payoff for an agent is given by

$$S_i^A = q_A \sum_{j \sim i} (1 - X_j) \quad \text{for strategy } A,$$

$$S_i^B = r + (q_B + u) \sum_{j \sim i} X_j \quad \text{for strategy } B. \tag{1}$$

We consider that q_A and q_B are fixed and correspond to the level of performance of the technologies A and B.

By (1), we have $S_i^B \leq S_i^A$ iff

$$r + (q_B + u) \sum_{j \sim i} X_j \leq q_A \sum_{j \sim i} (1 - X_j) \quad \Leftrightarrow \quad \sum_{j \sim i} X_j \leq \theta(d_i), \tag{2}$$

with $\theta(d) := \frac{q_A d - r}{q_A + q_B + u}$ and d_i is the degree (number of neighbours) of i. We now explain the dynamics of our model for the spread of strategy B in the network as time t evolves. We consider a fixed network G (not evolving in time) and let all agents play A for $t < 0$. At time $t = 0$, some agents are forced to strategy B. These agents will always play strategy B, hence the dynamics described below does not apply to these initially forced agents. We encode the initial population forced to strategy B by a vector χ, where $\chi_i = 1$ if agent i is forced to B and $\chi_i = 0$ otherwise. We will assume that the vector $\chi = (\chi_i)_{i \in V}$ is a sequence of i.i.d. Bernoulli random variables with parameter α.

3   Continuous-Time Dynamic

We first consider the following continuous version of the contagion model. Assume that each non infected node i updates its state at rate 1 and it holds on to strategy A if (2) is satisfied and switches to B if $\sum_{j \sim i} X_j(t) > \theta(d_i)$. The state at time t is represented by a vector $X(t)$. Denote by A the adjacency matrix of the graph G and let $\lambda_1(A)$ the spectral radius of A, namely, its largest eigenvalue and by d_{min} the minimum degree of graph G. In addition we will assume that the graph is connected so that $d_{min} \geq 1$ and $\lambda_1(A)$ has multiplicity one. Therefore we have, $X_i(0) = \chi_i$, for all $i \in V$, and

$$X_i : 0 \to 1 \quad \text{at rate } \mathbb{1}\left(\sum_j A_{ij} X_j(t) > \theta(d_i)\right).$$

Note that $\mathbb{1}\left(\sum_j A_{ij} X_j(t) > \theta(d_i)\right) \leq \frac{\sum_j A_{ij} X_j(t)}{\theta(d_i)}$. We now consider the continuous time Markov process $Z(t) = (Z_i(t))_{i \in V}$, with $Z(0) = X(0)$, and transition rate:

$$Z_i : k \to k + 1 \quad \text{at rate } \frac{\sum_{j=1}^n A_{ij} Z_j(t)}{\theta(d_{min})},$$

standard coupling arguments yield $X(t) \leq_{st} Z(t)$ for all $t \geq 0$, where $X \leq_{st} Z$ denotes that Z stochastically dominates X. This implies that $\sum_{i=1}^n \mathbb{E}(X_i(t)) \leq \sum_{i=1}^n \mathbb{E}(Z_i(t))$. Moreover, the transition rates of the process Z are such that

$$\frac{d\mathbb{E}[Z(t)]}{dt} = \frac{A}{\theta(d_{min})}\mathbb{E}[Z(t)],$$

Hence

$$\mathbb{E}[Z(t)] = e^{\frac{t}{\theta(d_{min})}A}\,\mathbb{E}[Z(0)]. \qquad (3)$$

Using Cauchy-Schwartz inequality, we obtain that $\sum_{i=1}^{n}\mathbb{E}(Z_i(t)) \leq ||\mathbb{E}(Z(t))||_2||1||_2$. Combining this with (3), we have that

Theorem 1. *Let $\beta(t)$ be the proportion of nodes that opted for strategy B by time t. Then*

$$\beta(t) := \frac{\sum_{i=1}^{n}\mathbb{E}(X_i(t))}{n} \leq \alpha e^{\frac{\lambda_1(A)}{\theta(d_{min})}t}.$$

Moreover if the G is a regular graph with degree Δ, then, using the spectral decomposition of the matrix $e^{\frac{t}{\theta(\Delta)}A}$, we have that

$$\beta(t) \leq \frac{\alpha}{\Delta}e^{\frac{\Delta}{\theta(\Delta)}t}$$

The above result states that the number of nodes that have adopted B increases at most exponentially in time and that the speed is given by $\frac{\lambda_1(A)}{\theta(d_{min})}$. Similar results have been found in [4] in the case of the Susceptible-Infected-Susceptible (SIS) epidemic.

As a matter of example for Erdös-Rényi graphs $G(n,p)$ with parameters n and p in the regime $\log(n) << np$, Theorem 1 yields $\beta(t) \leq \alpha e^{\frac{np}{\theta(np)}t}$ with high probability.

In the next section, we describe a discrete-time version of our contagion model for which we derive more accurate results for the proportion of B-adopters and illustrate the coexistence of the two strategies.

4   Discrete-Time Dynamic

The state of the network at time t is described by the vector $(X_i(t))_{i \in V}$, $t \in \mathbb{N}$. We have $X_i(0) = \chi_i$ and $X_i(t) \geq \chi_i$. Then at each time step $t \geq 1$, each agent applies the best-response update: if $S_i^B > S_i^A$ then he chooses B and if not then he chooses A. It is readily seen that

$$1 - X_i(t+1) = (1 - \chi_i)\mathbf{1}\left(S_i^B(t) \leq S_i^A(t)\right). \qquad (4)$$

4.1   Diffusion Process on the Infinite Regular Tree

Let $T(\Delta)$ be an infinite Δ-regular tree with nodes $\emptyset, 1, 2, \ldots$, with a fixed root $\emptyset$. For a node i, we denote by $\text{gen}(i) \in \mathbb{N}$ the generation of i, i.e. the length of the minimal path from $\emptyset$ to i. Also we denote $i \to j$ if i belongs to the children of j, i.e. $\text{gen}(i) = \text{gen}(j) + 1$ and j is on the minimal path from $\emptyset$ to i. For an edge (i, j) with $i \rightarrow j$, we denote by $T_{i \rightarrow j}$ the sub-tree of T with root i obtained by the deletion of edge (i, j) from T.

For a given vector χ, we say that node $i \neq \emptyset$ is infected from $T_{i \rightarrow j}$ if the node i switches to B in $T_{i \rightarrow j} \bigcup \{(i, j)\}$ with the same vector χ for $T_{i \rightarrow j}$ and the strategy A for j. We denote by $Y_i(t)$ the corresponding indicator function with value 1 if i is infected from $T_{i \rightarrow j}$ at time t and 0 otherwise.

Proposition 1. *We have*

$$1 - X_\emptyset(t+1) = (1 - \chi_\emptyset) \mathbb{1} \left(\sum_{i \sim \emptyset} Y_i(t) \leq \theta(\Delta) \right). \tag{5}$$

The representation (5) is crucial to our analysis. In fact, thanks to the tree structure, the random variables $(Y_i(t), i \sim \emptyset)$ are independent of each other and identically distributed. More precisely, a simple induction shows that (4) becomes, for $i \neq \emptyset$:

$$1 - Y_i(t+1) = (1 - \chi_i) \mathbb{1} \left(\sum_{j \rightarrow i} Y_j(t) \leq \theta(\Delta) \right). \tag{6}$$

Note that (6) allows to compute all the $Y_i(t)$ recursively, starting with $Y_i(0) = \chi_i$. It is then easy to compute their distribution from (6). We summarize this result in the next proposition.

Proposition 2. *For t fixed, the sequence $(Y_i(t), i \sim \emptyset)$ is a sequence of i.i.d. Bernoulli random variables with parameter $h(t)$ given by $h(0) = \alpha$ and, for $t \geq 0$,*

$$h(t+1) = \mathbb{P}(Y_i(t+1) = 1) = 1 - (1 - \alpha) g_{\Delta - 1, \theta(\Delta)}(h(t)),$$

where $g_{k,s}(x) = \mathbb{P}(Bin(k, x) \leq s)$. $Bin(k, x)$ corresponds to the binomial distribution with parameters k and x.

Combining Propositions 1 and 2, we obtain that

Theorem 2. *$X_\emptyset(t)$ is a Bernoulli random variable with parameter $\tilde{h}(t)$ given by*

$$\mathbb{P}(X_\emptyset(t+1) = 1) = \tilde{h}(t+1) = 1 - (1 - \alpha) g_{\Delta, \theta(\Delta)}(h(t)). \tag{7}$$

Moreover let h^* the smallest solution of the following fixed point equation

$$h = 1 - (1 - \alpha) g_{\Delta - 1, \theta(\Delta)}(h). \tag{8}$$

Suppose $0 \leq \theta(\Delta) < \Delta - 2$. There exists $\alpha_{crit} < 1$ such that for all $\alpha > \alpha_{crit}$, the fixed point equation (8) has a unique solution $h^* = 1$ and for all $\alpha < \alpha_{crit}$ it has three solutions $h^* < h^{**} < h^{***} = 1$.

4.2   Random Regular Graphs

We now come back to the process $(X_i^{(n)}(t))_{i\in V}$ on $G_\Delta^{(n)}$, a random $\Delta-$regular graph, satisfying (4). Given $d \geq 1$, let $N(i, d, G_\Delta^{(n)})$ be the set of vertices of $G_\Delta^{(n)}$ that are at a distance at most d from $i \in G_\Delta^{(n)}$. A depth-d Δ-regular tree $T(\Delta, d)$ is the restriction of $T(\Delta)$ to nodes i with gen$(i) \leq d$. A simple induction on t shows that $X_i^{(n)}(t)$ is determined by the $\{\chi_j, j \in N(i, t, G_\Delta^{(n)})\}$. Using the following convergence [6]: for any fixed $d \geq 1$, we have as n tends to infinity $N(0, d, G_\Delta^n) \xrightarrow{d} T(\Delta, d)$, we have $X_0^{(n)}(t) \xrightarrow{d} X_\emptyset(t)$ as n tends to infinity. Therefore the process defined on the tree in Section 4.1 is a good approximation of the real process. Hence,

Proposition 3. *For any fixed $t \geq 0$, we have*

$$\lim_{n\to\infty} \mathbb{E}\left[X_i^{(n)}(t)\right] = \tilde{h}(t) \tag{9}$$

where $\tilde{h}(t)$ is defined in (7).

Let $\beta^{(n)}(t)$ be the proportion of agents choosing B at time t: $\beta^{(n)}(t) = \sum_i X_i^{(n)}(t)/n$. We have as $n \to \infty$,

$$\mathbb{E}\left[\beta^{(n)}(t)\right] = \mathbb{E}\left[X_i^{(n)}(t)\right] \to \tilde{h}(t). \tag{10}$$

The final proportion of agents choosing B is $\beta^{(n)} = \lim_{t\to\infty} \beta^{(n)}(t)$.

Proposition 4. *We have*

$$\lim_{n\to\infty} \mathbb{E}\left[\beta^{(n)}\right] = \tilde{h}, \tag{11}$$

in particular, for $\alpha \geq \alpha_{crit}$, we have

$$\lim_{n\to\infty} \mathbb{E}\left[\beta^{(n)}\right] = 1. \tag{12}$$

The interchange of limits in t and n needs a proper mathematical proof. This has been done in [1] and our statement follows from their Theorem 1. For Δ-regular graphs, bootstrap percolation is equivalent to our model. It is noticed in [1] that the critical value on the Δ-regular random graph turns out to be the same as that on the Δ-tree, although the proof goes along a quite different route.

4.3   Extensions: Random Networks and Linear Threshold Model

The study of games on networks has attracted a lot of attention during the recent year. We refer to [3] for the description of a general framework and references to the economic literature. An attractive way to model the social network consists in using a mean field approach to study the random network. The main advantage of this approach is its simplicity. A random network is characterized by its connectivity distribution: the fraction of agents in the population with d neighbors is described by the degree distribution $P(d)$. Such a model is analyzed in [5], where the dynamic of the diffusion process is also considered.

In this section, we show how our approach extends to random networks and to the linear the linear threshold model (see [9], [8] for a rigorous proof). Let us assume that the graph $G^{(n)}$ is defined via its degree sequence $(D_i)_{i \in V}$ which is i.i.d. distributed according to D. Such graphs can be generated using the configuration model [13]. Let $L_n = \sum_{k=1}^{n} D_k$, where $L_n/2$ is the number of edges in the graph. The underlying tree rooted at node i can be described by a branching process with the offspring distribution of the root given by D_i. Besides the subsequent generations have offspring distribution

$$p_j^{(n)} = \sum_{k=1}^{n} \mathbb{1}(D_k = j+1) \, \frac{D_k}{L_n} .$$

If the degree sequence is such that $\mathbb{E}[D^2]$ is finite then, by the strong law of large numbers

$$\lim_{n \to \infty} p_j^{(n)} = \frac{(j+1)\mathbb{P}(D_1 = j+1)}{\mathbb{E}D} , \quad a.s.$$

Let $\tilde{P}(j) = \frac{jP(j)}{\sum_j jP(j)}$ be the (asymptotic) probability that an edge points to a node with degree j. Then for any fixed d, the neighbourhood of radius d about node 0, $N(0, d, G^{(n)})$ converges in distribution as n tends to infinity to a depth-d a Galton-Watson tree with a root which has offspring distribution P and all other nodes have offspring distribution P^* given by $P^*(j-1) = \tilde{P}(j)$ for all $j \geq 1$. Thus the associated fixed point equation is:

$$h^* = 1 - (1-\alpha) \sum_j P^*(j) g_{j, \theta(j+1)}(h^*), \tag{13}$$

and we have $\lim_{t \to \infty} \beta(t) = \tilde{h}$ given by

$$\tilde{h} = 1 - (1-\alpha) \sum_j P(j) g_{j, \theta(j)}(h^*). \tag{14}$$

As a matter of example, for Erdős-Rényi graphs, the fixed point equation associated with our model is given by (13) and (14) with $P(j) = P^*(j) = e^{-\lambda} \frac{\lambda^j}{j!}$.

Besides the methods developed in this paper are not restricted to this particular model. We consider the general threshold model [7]. We have a non-negative random weight W_{ij} on each edge, indicating the influence that i exerts on j. We consider the symmetric case where $W_{ij} = W_{ji}$ and we assume W_{ij} are i.i.d with distribution function W. Each node i has an arbitrary function f_i defined on subsets of its neighbours set N_i: for any set of neighbours $X \subseteq N_i$, there is a value $f_i(X)$ between 0 and 1 which is monotone in the sense that if $X \subseteq Y$, then $f_i(X) \leq f_i(Y)$. This node chooses a threshold θ_i at random from $[0, 1]$ and at time step $t + 1$ it becomes active, it plays B, if its set of currently active neighbours $N_i^B(t)$ satisfies $f_i(N_i^B(t)) > \theta_i$.

5   Conclusion and Future Work

In this paper we presented two models of marketing wherein individuals, represented by a graph structure, receive payoffs to entice them to adopt a strategy that is different from their initial choice. To this end we initially force a small proportion of nodes to opt for the new strategy and then use an economic model that accounts for the cascading dynamic of adoption. We analyze the evolution of the proportion of agents that switch to the new strategy over time. First, the implications of our results concern marketing strategies in online social networks. More precisely, let $\alpha = \frac{\sum_i X_i}{n}$ be the proportion of forced agents and let $M_1(\alpha)$ the price incurred to force the initial agents. Typically if there is a fixed cost per agent, say c, we could take $M_1(\alpha) = c\alpha$. Let $\beta(t)$ be the proportion of agents choosing B at time t: $\beta(t) = \frac{\sum_i X_i(t)}{n}$. We have $\gamma(t) = \beta(t) - \alpha \geq 0$ which corresponds to the proportion of agents choosing B without being initially forced. We denote by $M_2(\gamma(t))$ the price incurred by the rebates until time t. We typically take $M_2(\gamma) = r\gamma$. Let $\delta(t)$ be the proportion of edges $B - B$ at time t. We denote by $M_3(\delta(t))$ the price incurred by the marketing of edges until time t. We typically take $M_3(\delta) = u\delta$. Hence the total price of the marketing strategy at time t is given by $M(t) = M_1(\alpha) + M_2(\gamma(t)) + M_3(\delta(t))$. One can compute the quantities $\gamma(t)$ and $\delta(t)$ in function of α, r and u. This opens the possibilities of doing an optimal control of the marketing policy.

Finally we remark that the marketing problem that we considered in this paper is just one application of our method. Our approach can indeed be adapted to the analysis of the dissemination of new versions of existing protocols, voting protocols through simple majority rules, i.e., $\theta(d) = \frac{d}{2}$ and distributed digital preservation systems [11].

References

1. Balogh, J., Pittel, B.G.: Bootstrap percolation on the random regular graph. Random Structures Algorithms 30(1-2), 257–286 (2007)
2. Domingos, P., Richardson, M.: Mining the network value of customers. In: KDD 2001: Proceedings of the seventh ACM SIGKDD international conference on Knowledge discovery and data mining, pp. 57–66. ACM Press, New York (2001)
3. Galeotti, A., Goyal, S., Jackson, M.O., Vega-Redondo, F., Yariv, L.: Network games (2007)
4. Ganesh, A., Massoulié, L., Towsley, D.: The effect of network topology on the spread of epidemics. In: Proceedings IEEE Infocom (2005)
5. Jackson, M.O., Yariv, L.: Diffusion of behavior and equilibrium properties in network games. American Economic Review (Papers and Proceedings) 97(2), 92–98 (2007)
6. Janson, S., Łuczak, T., Rucinski, A.: Random graphs. Wiley-Interscience Series in Discrete Mathematics and Optimization. Wiley Interscience, New York (2000)
7. Kleinberg, J.: Cascading behavior in networks: Algorithmic and economic issues. In: Nisan, N., Roughgarden, T., Tardos, E., Vazirani, V. (eds.) Algorithmic Game Theory. Cambridge University Press, Cambridge (2007)

8. Lelarge, M.: Diffusion and cascading behavior in random networks (in preparation)
9. Lelarge, M.: Diffusion of innovations on random networks: Understanding the chasm. In: Papadimitriou, C.H., Zhang, S. (eds.) WINE 2008. LNCS, vol. 5385, pp. 178–185. Springer, Heidelberg (2008)
10. Mahajan, V., Muller, E., Bass, F.M.: New product diffusion models in marketing: A review and directions for research. Journal of Marketing 54(1), 1–26 (1990)
11. Maniatis, P., Roussopoulos, M., Giuli, T., Rosenthal, D.S.H., Baker, M.: The lockss peer-to-peer digital preservation system. ACM Transactions on Computer Systems 23 (2005)
12. Morris, S.: Contagion. Rev. Econom. Stud. 67(1), 57–78 (2000)
13. van der Hofstad, R.: Random graphs and complex networks (2008), http://www.win.tue.nl/~rhofstad/NotesRGCN2008.pdf
14. Watts, D.: A simple model of global cascades on random networks. Proceedings of the National Academy of Science 99, 5766–5771 (2002)

A Stochastic Epidemiological Model and a Deterministic Limit for BitTorrent-Like Peer-to-Peer File-Sharing Networks

George Kesidis[1,*], Takis Konstantopoulos[2,**], and Perla Sousi[3]

[1] CSE and EE Departments, Pennsylvania State University,
University Park, PA 16802, USA
gik2@psu.edu

[2] School of Mathematical & Computer Sciences and the Maxwell Institute for
Mathematical Sciences, Heriot-Watt University, Edinburgh EH14 4AS, UK
takis@ma.hw.ac.uk

[3] Statistical Laboratory, Centre for Mathematical Sciences, University of Cambridge,
Wilberforce Road, Cambridge CB3 0WB, UK
P.Sousi@statslab.cam.ac.uk

Abstract. We propose a stochastic model for a file-sharing peer-to-peer network which resembles the popular BitTorrent system: large files are split into chunks and a peer can download or swap from another peer only one chunk at a time. We exhibit the fluid and diffusion limits of a scaled Markov model of this system and look at possible uses of them to draw practical conclusions.

1 Introduction

Peer-to-peer (p2p) activity continues to represent a very significant fraction of overall Internet traffic, 44% by one recent account [4]. BitTorrent [1,2,8,21,18,9,19] is a widely deployed p2p file-sharing network which has recently played a significant role in the network neutrality debate. Under BitTorrent, peers join "swarms" (or "torrents") where each swarm corresponds to a specific data object (file). The process of finding the peers in a given swarm to connect to is typically facilitated through a centralised "tracker". Recently, a trackerless BitTorrent client has been introduced that uses distributed hashing for query resolution [16].

For file sharing, a peer is typically uploads upload pieces ("chunks") of the file to other peers in the swarm while downloading his/her missing chunks from them. This chunk swapping constitutes a transaction-by-transaction incentive for peers to cooperate (i.e., trading rather than simply download) to disseminate data objects. Large files may be segmented into several hundred chunks, all of which the peers of the corresponding warm must collect, and in the process disseminate their own chunks before they can reconstitute the desired file and possibly leave the file's swarm.

---

* Supported by an NSF grant.
** Corresponding author; supported by an EPSRC grant.

E. Altman and A. Chaintreau (Eds.): NET-COOP 2008, LNCS 5425, pp. 26–36, 2009.

In addition to the framework in which data objects are segmented into chunks to promote cooperation through swapping, there is a system whereby the rate at which chunks are uploaded is assessed for any given transaction, and peers that allocate inadequate bandwidth for uploading may be "choked" [14,17]. Choking may also be applied to peers who, by employing multiple identities (sybils), abuse BitTorrent's system of allowing newly arrived peers to a swarm to just download a few chunks (as they clearly cannot trade what they simply do not as yet possess). BitTorrent can also rehabilitate peers by (optimistically) unchoking them. In the following, we do not directly consider upload bandwidth and related choking issues.

In this paper, we motivate a deterministic epidemiological model of file dissemination for peer-to-peer file-sharing networks that employ BitTorrent-like incentives, a generalisation of that given in [10,11]. Our model is different from those explored in [15,21,18] for BitTorrent, and we compute different quantities of interest. Our epidemiological framework, similar to that we used for the spread of multi-stage worms [12], could also be adapted for network coding systems. In [9], the authors propose a "fluid" model of a single torrent/swarm (as we do in the following) and fit it to (transient) data drawn from aggregate swarms. The connection to branching process models [21,8] is simply that ours only tracks the number of active peers who possess or demand the file under consideration, i.e., a single swarm. Though our model is significantly simpler than that of prior work, it is derived directly from an intuitive transaction-by-transaction Markov process modelling file-dissemination of the p2p network and its numerical solutions clearly demonstrate the effectiveness of the aforementioned incentives. A basic assumption in the following is that peers do not distribute bogus files (or file chunks) [20].

2   The Stochastic Model

A file is represented as a set F of size n, the elements of which are called chunks. Consider a large networked "swarm" of N nodes called peers. Each peer possesses a certain (possibly empty) subset A of F. As time goes by, this peer interacts with other peers, the goal being to enlarge his set A until, eventually, the peer manages to collect all n chunks of F. The interaction between peers can either be a download or a swap; in both cases, chunks are being copied from peer to peer and are assumed never lost. A peer will stay in the network as long as he does not possess all chunks. After collecting everything, sooner or later a peer departs or switches off. By splitting the desired file into many chunks we give incentives to the peers to remain active in the swarm for long time during which other peers will take advantage of their possessions.

2.1   Possible Interactions

We here describe how two peers, labelled A, B, interact. The following types of interactions are possible:

1. **Download:** Peer A downloads a chunk i from B. This is possible only if A is a strict subset of B. If $i \in B$ then, after the downloading A becomes $A' = A \cup \{i\}$ and but B remains B because it since it gains nothing from A. Denote this interaction by: $(A \leftarrow B) \rightsquigarrow (A', B)$. The symbol on the left is supposed to show the type of interaction and the labels before it, while the symbol on the right shows the labels after the interaction.

2. **Swap:** Peer A swaps with peer B. In other words, A gets a chunk j from B and B gets a chunk i from A. It is required that j is not an element of A and i not an element of B. We denote this interaction by $(A \leftrightarrows B) \rightsquigarrow (A', B')$, where $A' = A \cup \{j\}$, $B' = B \cup \{i\}$. We thus need $A \setminus B \neq \varnothing$ and $B \setminus A \neq \varnothing$.

2.2 Notation

The set of all combinations of n chunks, which partition F, is denoted by $\mathscr{P}(F)$, where $|\mathscr{P}(F)| = 2^n$ and the empty set is included. We write $A \subset B$ (respectively, $A \subsetneq B$) when A is a subset (respectively, strict subset) of B. We (unconventionally) write

$$A \sqsubset A' \text{ when } A \subset A' \text{ and } |A' - A| = 1.$$

If $A \cap B = \varnothing$, we use $A + B$ instead of $A \cup B$; if $B = \{b\}$ is a singleton, we often write $A + b$ instead of $A + \{b\}$. If $A \subset B$ we use $B - A$ instead of $B \setminus A$. We say that

$$A \text{ relates to } B \text{ (and write } A \sim B \text{) when } A \subset B \text{ or } B \subset A;$$

if this is not the case, we write $A \not\sim B$. Note that $A \not\sim B$ if and only if two peers labelled A, B can swap chunks. The space of functions (vectors) from $\mathscr{P}(F)$ into $\mathbb{Z}_+$ is denoted by $\mathbb{Z}_+^{\mathscr{P}(F)}$. The stochastic model will take values in this space. The deterministic model will evolve in $\mathbb{R}_+^{\mathscr{P}(F)}$. We let $e_A \in \mathbb{Z}_+^{\mathscr{P}(F)}$ be the vector with coordinates

$$e_A^B := \mathbf{1}(A = B), \quad B \in \mathscr{P}(F).$$

For $x \in \mathbb{Z}_+^{\mathscr{P}(F)}$ or $\mathbb{R}_+^{\mathscr{P}(F)}$ we let $|x| := \sum_{A \in \mathscr{P}(F)} |x^A|$.

2.3 Defining the Rates of Individual Interactions

We follow the logic of stochastic modelling of chemical reactions or epidemics and assume that the chance of a particular interaction occurring in a short interval of time is proportional to the number of ways of selecting the peers needed for this interaction [13]. Accordingly, the interaction rates *must* be given by the formulae described below.

Consider first finding the rate of a download $A \leftarrow B$, where $A \subsetneq B$, when the state of the system is $x \in \mathbb{Z}_+^{\mathscr{P}(F)}$. There are x^A peers labelled A and x^B labelled B. We can choose them in $x^A x^B$ ways. Thus the rate of a download $A \leftarrow B$ that results into A getting *some* chunk from B should be proportional to $x^A x^B$. However, we are interested in the rate of the *specific* interaction $(A \leftarrow B) \rightsquigarrow (A', B)$, that turns A into a specific set A' differing from A by one single chunk

$(A \sqsubset A')$; there are $|B - A|$ chunks that A can download from B; the chance that picking one of them is $1/|B - A|$. Thus we have:

$$(DR) \quad \begin{cases} \text{the rate of the download } (A \leftarrow B) \rightsquigarrow (A', B) \text{ equals } \beta x^A \dfrac{x^B}{|B - A|}, \\ \text{as long as } A \sqsubset A' \subset B, \end{cases}$$

where $\beta > 0$.

Consider next a swap $A \leftrightarrows B$ and assume the state is x. Picking two peers labelled A and B (provided that $A \not\sim B$) from the population is done in $x^A x^B$ ways. Thus the rate of a swap $A \leftrightarrows B$ is proportional to $x^A x^B$. So if we *fix* two chunks $i \in A \setminus B, j \in B \setminus A$ and specify that $A' = A + j, B' = B + i$, then the chance of picking i from $A \setminus B$ and j from $B \setminus A$ is $1/|A \setminus B||B \setminus A|$. Thus,

$$(SR) \quad \begin{cases} \text{the rate of the swap } (A \leftrightarrows B) \rightsquigarrow (A', B') \text{ equals } \gamma \dfrac{x^A x^B}{|A \setminus B||B \setminus A|}, \\ \text{a long as } A \sqsubset A', \quad B \sqsubset B', \quad A' - A \subset B, \quad B' - B \subset A, \end{cases}$$

where $\gamma > 0$.

2.4   Deriving the Markov Chain Rates

Having defined the rates of each individual interaction we can easily define rates $q(x, y)$ of a Markov chain in continuous time and state space $\mathbb{Z}_+^{\mathscr{P}(F)}$ as follows.

Define functions $\lambda_{A,A'}, \mu_{A,B} : \mathbb{R}^{\mathscr{P}(F)} \to \mathbb{R}$ by:

$$\lambda_{A,A'}(x) := \left[\beta x^A \sum_{C : C \supset A'} \frac{x^C}{|C - A|} \right] \mathbf{1}(A \sqsubset A') \tag{1a}$$

$$\mu_{A,B}(x) := \gamma \frac{x^A x^B}{|A \setminus B||B \setminus A|} \mathbf{1}(A \not\sim B). \tag{1b}$$

Consider also constants $\delta \geq 0$ and $\alpha^A \geq 0$ for $A \in \mathscr{P}(F)$, i.e., $\alpha \in \mathbb{R}_+^{\mathscr{P}(F)}$. The transition rates of the closed conservative Markov chain are given by:

$$q(x, y) := \begin{cases} \lambda_{A,A'}(x), & \text{if } y = x - e_A + e_{A'} \\ \mu_{A,B}(x), & \text{if } \begin{cases} y = x - e_A - e_B + e_{A'} + e_{B'} \\ A \sqsubset A', B \sqsubset B', A' - A \subset B, B' - B \subset A, \end{cases} \\ \alpha^A & \text{if } y = x + e_A \\ \delta x^F & \text{if } y = x - e_F \\ 0, & \text{for any other value of } y \neq x, \end{cases} \tag{2}$$

where x ranges in $\mathbb{Z}_+^{\mathscr{P}(F)}$.

A little justification of the first two cases is needed: that $q(x, x - e_A - e_B + e_{A'} + e_{B'}) = \mu_{A,B}(x)$ is straightforward. It corresponds to a swap, which is only possible when $A \sqsubset A', B \sqsubset B', A' - A \subset B, B' - B \subset A$. The swap rate was defined by (SR). To see that $q(x, x - e_A + e_{A'}) = \lambda_{A,A'}(x)$ we observe that a peer labelled A can change its label to $A' \sqsupset A$ by downloading a chunk from some set C that contains A', so we sum the rates (DR) over all these possible individual interactions to obtain the first line in (2). We can think of having Poisson process of arrivals of new peers at rate $|\alpha|$, and that each arriving peer is labelled A with probability $\alpha^A/|\alpha|$. Peers can depart, by definition, only when they are labelled F and it takes an exponentially distributed amount of time (with mean $1/\delta$) for a departure to occur. Thus, $q(x, x - e_F) = \delta x^F$. We shall let Q denote the generator of the chain, i.e. $Qf(x) = \sum_y (f(y) - f(x))q(x,y)$, when f is an appropriate functional of the state space.

Definition 1 (BITTORRENT $[x_0, n, \alpha, \beta, \gamma, \delta]$). *Given $x_0 \in \mathbb{Z}_+^{\mathscr{P}(F)}$ (initial configuration), $n = |F| \in \mathbb{N}$ (number of chunks), $\alpha \in \mathbb{R}_+^{\mathscr{P}(F)}$ (arrival rates), $\beta > 0$ (download rate), $\gamma \geq 0$ (swap rate), $\delta \geq 0$ (departure rate) we let BITTORRENT $[x_0, n, \alpha, \beta, \gamma, \delta]$ be a Markov chain $(X_t, t \geq 0)$ with transition rates (2) and $X_0 = x_0$. We say that the chain (network) is $\underline{open}$ if $\alpha^A > 0$ for at least one A and $\delta > 0$; it is $\underline{closed}$ if $\alpha^A = 0$ for all A; it is $\underline{conservative}$ if it is closed and $\delta = 0$; it is $\underline{dissipative}$ if it is closed and $\delta > 0$.*

In a conservative network, we have $q(x,y) = 0$ if $|y| \neq |x|$ and so $|X_t| = |X_0|$ for all $t \geq 0$. Here, the actual state space is the simplex $\{x \in \mathbb{Z}_+^{\mathscr{P}(F)} : |x| = N\}$, where $N = |X_0|$. It is easy to see that the state e_F is reachable from any other state, but all rates out of e_F are zero. Hence a conservative network has e_F as a single absorbing state.

In a dissipative network, we have $|X_t| \leq |X_0|$ for all $t \geq 0$. Here the state space is $\{x \in \mathbb{Z}_+^{\mathscr{P}(F)} : |x| \leq N\}$, where $N = |X_0|$. It can be seen that a dissipative network has many absorbing points.

In an open network, there are no absorbing points. On the other hand, one may wonder if certain components can escape to infinity. This is not the case:

Lemma 1. *If $\alpha^F > 0$ then the open* BITTORRENT $[x, n, \beta, \gamma, \alpha, \delta]$ *is positive recurrent Markov chain.*

Proof. (sketch) If $\alpha^F > 0$, $\delta > 0$ the Markov chain is irreducible. The remainder of the proof is based on a the construction of a simple Lyapunov function: $V(x) := |x|$, for which it can be shown that there is a bounded set of states K such that $\sup_{x \notin K}(QV)(x) < 0$. Perhaps the easiest way to see this is by appealing to the stability of the corresponding ODE limit; see Theorem 1 below and [7]. $\qquad\square$

3 Macroscopic Description: Fluid Limit and Diffusion Approximation

Analysing the Markov chain in its original form is complicated. We thus resort to a first-order approximation by an ordinary differential equation (ODE). Let $v(x)$ be the vector field on $\mathbb{R}_+^{\mathscr{P}(F)}$ with components $v^A(x)$ defined by

$$v^A(x) = \alpha^A - x^A \left(\beta \varphi_d^A(x) + \gamma \varphi_s^A(x) \right)$$

$$+ \beta \sum_{B:A \subset B} \frac{\psi_d^A(x) x^B}{1 + |B \setminus A|} + \gamma \sum_{B:A \not\subset B} \frac{\psi_s^{A,B}(x) x^B}{1 + |B \setminus A|} - \delta x^F \mathbf{1}(A = F), \quad (3)$$

where

$$\varphi_d^A(x) := \sum_{B \supset A} x^B, \quad \varphi_s^A(x) := \sum_{B \not\supset A} x^B$$

$$\psi_d^A(x) := \sum_{a \in A} x^{A-a}, \quad \psi_s^{A,B}(x) := \sum_{a \in A \cap B} x^{A-a} \quad (4)$$

Consider the differential equation

$$\dot{x} = v(x) \text{ with initial condition } x_0. \quad (5)$$

Consider the sequence of stochastic models $\texttt{BITTORRENT}\,[X_{N,0}, n, N\alpha, \frac{\beta}{N}, \frac{\gamma}{N}, \delta]$ for $N \in \mathbb{N}$ and let $X_{N,t}$ be the corresponding jump Markov chain.

Theorem 1. *There is a has a unique smooth (analytic) solution to (5), denoted by x_t for $t \geq 0$. Also, if there is an $x_0 \in \mathbb{R}_+^{\mathscr{P}(F)}$ such that $X_{N,0}/N \to x_0$ as $N \to \infty$, then for any $T, \varepsilon > 0$,*

$$\lim_{N \to \infty} P\left(\sup_{0 \leq t \leq T} |N^{-1} X_{N,t} - x_t| > \varepsilon \right) = 0.$$

Proof. See [11].

Next, let

$$Y_{N,t} := \sqrt{N}(X_{N,t}/N - x_t).$$

For each $y \in \mathbb{Z}_+^{\mathscr{P}(F)}$ let W_y be a standard one-dimensional Brownian motion; suppose that these Brownian motions are independent over y. Define the (time-inhomogeneous) Gaussian diffusion process Y by

$$dY(t) = \sum_y (y - x_t) \sqrt{q(x_t, x_t + y)} dW_y(t) + Dv(x_t) Y(t) dt,$$

where $Dv(x)$ is the matrix of partial derivatives of $v(x)$. Due to the form the rates (2), the first sum ranges over finitely many y and so only finitely many Brownian motions are needed.

Theorem 2. *If $\sqrt{N}(X_{N,t}/N - x_0) \to 0$ as $N \to \infty$, where $x_0 \in \mathbb{R}_+^{\mathscr{P}(F)}$, then the law of Y_N (as a sequence of probability measures in $D[0, \infty)$ with the topology of uniform convergence on compacta) converges weakly to the law of Y.*

Proof. We refer to [13] for the relevant arguments.

4   Examples

Suppose that F consists of $n = 2$ chunks. The limiting ODE is easily found to be:

$$\dot{x}^{\varnothing} = \alpha^{\varnothing} - \beta x^{\varnothing}(x^1 + x^2 + x^{12})$$
$$\dot{x}^1 = \alpha^1 - x^1(\beta x^{12} + \gamma x^2) + \beta x^{\varnothing}(x^1 + \tfrac{1}{2}x^{12})$$
$$\dot{x}^2 = \alpha^2 - x^2(\beta x^{12} + \gamma x^1) + \beta x^{\varnothing}(x^2 + \tfrac{1}{2}x^{12})$$
$$\dot{x}^{12} = \alpha^{12} + \beta(x^1 + x^2)x^{12} + 2\gamma x^1 x^2 - \delta x^{12}.$$

We look at its behaviour in three cases. To make things easier, assume that $\gamma = 0$.

4.1   Closed Conservative System: $\alpha^1 = \alpha^2 = \alpha^{12} = 0, \delta = 0$

Letting $x = x^{\varnothing}, u = x^1 + x^2, w = x^{12}$, assuming that $x + u + w = 1$, and eliminating the variable w we obtain

$$\dot{x} = -\beta x(1 - x)$$
$$\dot{u} = \beta u^2 - \beta u(1 - x) + \beta x(1 - x).$$

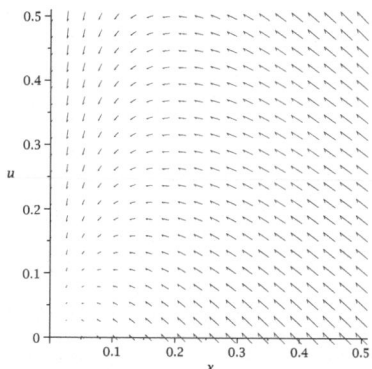

Fig. 1. Typical vector field plot for a closed conservative BitTorrent model

The vector field in the $x - u$ plane is depicted in Figure 1. The unique equilibrium point $x = 0, u = 0$ corresponds to $w = 1$, i.e. everybody possesses the full file. By solving the equation in x, substituting into the equation for u, we find the explicit solution

$$w_t = \frac{x_0 + (1 - x_0)e^{\beta t}}{x_0 + (1 - x_0)e^{\beta t} + x_0\beta t + (1 - w_0)w_0^{-1}},$$

from which one can estimate the time required for w to reach an ε-neighbourhood of the equilibrium, which can be turned into an estimate for the original stochastic system.

4.2   Closed Dissipative System: $\alpha^1 = \alpha^2 = \alpha^{12} = 0, \delta > 0$

With $x = x^\varnothing, u = x^1 + x^2, w = x^{12}$ as before, change the time variable to $s = \beta t$, let $\rho = \delta/\beta$, and write x' for dx/ds, to obtain:

$$x' = -x(u + w)$$
$$u' = -uw + x(u + w)$$
$$w' = uw - \rho w.$$

Assume $x_0 + u_0 + w_0 = 1$, so that $x_t + u_t + w_t < 1$ for all $t > 0$. We cannot eliminate the variable w now since there is no obvious conserved quantity, but we can study the equilibria of the system. It is easily seen that the only equilibria are of the form $(0, u, 0)$ which are unstable if $u > \rho$ and stable if $u < \rho$. In terms of the original variables, the stable equilibria are $(x^\varnothing, x^1, x^2, x^{12}) = (0, x^1, x^2, 0)$, $0 \leq x^1 + x^2 < \rho$. This is as expected: since there is no swapping ($\gamma = 0$), the system eventually settles to a situation where there are peers with label 1 and peers with label 2. Had γ been positive, x^1, x^2 could not have simultaneously been positive in equilibrium.

4.3   Open System: $\alpha^1 = \alpha^2 = 0, \alpha^{12} = \lambda > 0, \delta > 0$

Choosing variables appropriately, we have

$$x' = -x(u + w)$$
$$u' = -uw + x(u + w)$$
$$w' = \lambda + uw - \rho w.$$

The system eventually settles to the unique stable equilibrium $(x, u, w) = (0, 0, \lambda/\rho)$. It is easily seen that the eigenvalues of the differential of the vector field at this point are $-\lambda/\rho$ and $-\rho$ (the first one has algebraic multiplicity 1 but geometric multiplicity 2), and so there is no possibility of spiralling.

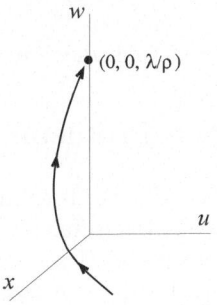

Fig. 2. Trajectory for open system

5 Performance Improvement in Presence of BitTorrent Incentives

To split or not to split? The answer is yes, but not always. We can address this question by looking at the behaviour of the ℓ_1-norm $|x^*|$ of the unique equilibrium point of an open system as the number of chunks (dimension) increases.

As an example, consider a situation where peers possessing nothing arrive at rate λ, download at rate β and depart at rate δ once they have the full file. This is described by

$$\dot{x}^\varnothing = \lambda - \beta x^\varnothing x^1$$
$$\dot{x}^1 = \beta x^\varnothing x^1 - \delta x^1.$$

The globally attracting stable equilibrium is given by $x^* = (\delta/\beta, \ \lambda/\delta)$.

Suppose now that we split into $n = 2$ chunks. Peers arrive and depart at the same rates but download at a rate $\widetilde{\beta} \geq \beta$ and swap at rate $\widetilde{\gamma}$. The new equilibrium is easily found to be

$$\widetilde{x}^* = \left(\frac{\delta}{\widetilde{\beta}} \Big(\frac{\delta}{\lambda} u + 1 \Big)^{-1}, \ \frac{u}{2}, \ \frac{u}{2}, \ \frac{\lambda}{\delta} \right),$$

where u is the positive number which solves

$$q(u) := u^2 + \frac{2\widetilde{\beta}\lambda}{\widetilde{\gamma}\delta} u - \frac{2\lambda}{\widetilde{\gamma}} = 0. \tag{6}$$

The following is shown in [11]:

Lemma 2

1. $\widetilde{x}^{*\varnothing} < x^{*\varnothing}$.
2. $|x^*| > |\widetilde{x}^*|$ if and only if $u < \widetilde{u}$, where $\widetilde{u}$ is the unique positive number which satisfies

$$\widetilde{q}(\widetilde{u}) := \widetilde{u}^2 - \Big(\frac{\delta}{\beta} - \frac{\lambda}{\delta} \Big) \widetilde{u} - \Big(\frac{\lambda}{\beta} - \frac{\lambda}{\widetilde{\beta}} \Big). \tag{7}$$

Furthermore, there exists a $\lambda_0 > 0$ such that for all $\lambda < \lambda_0$ (7) holds.

6 Conclusions and Open Problems

We proposed a stochastic model of a BitTorrent-like network and showed the existence of an ODE limit, along with a diffusion approximation. Several simulations [10] test the suitability of the model. Proofs of some of the results presented in this paper can be found in [11]. One can look at the ODE limit and, more specifically, its equilibria in order to obtain crude information about the stationary distribution of the original model. This requires proving a certain robustness result, as explained in [11, Sec. 5]. Rates of convergence to the ODE limit is another interesting open problem. Its solution requires estimating bounds on the vector field and its derivative. A criticism of the model is that the ODE lives in a high-dimensional space. One can reduce the dimension under some symmetry assumptions on the initial state and the parameters of the model [11, Sec. 7]. Another drawback is the Markovian assumption requires that times between transactions are exponential; this is clearly violated in practise (download times are typically heavy-tailed random variables). A quick fix of this problem is to represent the Markov process on a probability space supporting i.i.d. Poisson processes and then replace these by more general renewal processes.

References

1. BitTorrent, http://www.bittorrent.com
2. Cohen, B.: Incentives Build Robustness in BitTorrent. In: Workshop on Economics of Peer-to-Peer Systems, Berkeley, CA, USA (May 2003)
3. Daley, D.J., Gani, J.: Epidemic Modeling, an Introduction. Cambridge University Press, Cambridge (1999)
4. DC Info (Distributed Systems Newsletter) vol. XXII(8), June 30 (2008), http://www.dcia.info/news/#Newsletters
5. Ethier, S.N., Kurtz, T.G.: Markov Processes: Characterization and Convergence. Wiley, New York (1986)
6. Darling, R.W.R., Norris, J.R.: Differential equation approximations for Markov chains. Probability Surveys 5, 37–79 (2008)
7. Foss, S., Konstantopoulos, T.: An overview of some stochastic stability methods. Journal of the Operations Research Society of Japan 47, 275–303 (2004)
8. Ge, Z., Figueiredo, D.R., Jaiswal, S., Kurose, J., Towsley, D.: Modeling peer-to-peer file sharing systems. In: Proc. IEEE INFOCOM, San Francisco (April 2004)
9. Guo, L., Chen, S., Xiao, Z., Tan, E., Ding, X., Zhang, X.: Measurements, modeling and analysis of BitTorrent-like systems. In: Proc. Internet Measurement Conference (IMC) (October 2005)
10. Kesidis, G., Jin, Y., Mortazavi, B., Konstantopoulos, T.: An epidemiological model for file-sharing with BitTorrent-like incentives. In: Proc. IEEE GLOBECOM (November 2006)
11. Kesidis, G., Konstantopoulos, T., Sousi, P.: A stochastic epidemiological model and a deterministic limit for BitTorrent-like peer-to-peer file-sharing networks (November 2008), http://arxiv.org/abs/0811.1003
12. Kesidis, G., Vojnovic, M., Hamadeh, I., Jin, Y., Jiwasurat, S.: A Model of the Spread of Randomly Scanning Internet Worms that Saturate Access Links. In: ACM TOMACS (May 2008)
13. Kurtz, T.: Approximation of Population Processes. SIAM, Philadelphia (1981)
14. Legout, A., Liogkas, N., Kohler, E., Zhang, L.: Clustering and sharing incentives in BitTorrent systems. ACM SIGMETRICS Performance Evaluation Review 35(1) (June 2007)
15. Massoulié, L., Vojnovic, M.: Coupon replication systems. In: Proc. ACM SIGMETRICS, Banff, Alberta, Canada (2005)
16. Maymounkov, P., Mazieres, D.: Kademlia: a peer-to-peer information system based on the XOR metric. In: Proc. IPTPS, Cambridge, MA, USA (March 2002)
17. Mortazavi, B., Kesidis, G.: A peer-to-peer content-distribution game with a reputation-based incentive mechanism. In: Proc. IEEE Workshop on Information Theory and its Applications (ITA), UC San Diego (February 2006)

18. Qiu, D., Srikant, R.: Modeling and performance analysis of BitTorrent-like peer-to-peer networks. In: Proc. ACM SIGCOMM, Portland, Oregon (2004)
19. Turner, B.: Generalizing BitTorrent: how to build data exchange markets (and profit from them!) (January 2005),
    http://www.fractalscape.org/GeneralizingBitTorrent.htm
20. Walsh, K., Sirer, E.G.: Fighting peer-to-peer SPAM and decoys with object reputation. In: Proc. Workshop on Economics of Peer-to-Peer Systems (p2pecon), Philadelphia, PA (August 2005)
21. Yang, X., de Veciana, G.: Service capacity of peer to peer networks. In: Proc. IEEE INFOCOM, San Francisco (2004)

Ant Colony Optimization Algorithms for Shortest Path Problems

Sudha Rani Kolavali and Shalabh Bhatnagar*

Indian Institute of Science, Computer Science and Automation, Bangalore
sudha@csa.iisc.ernet.in, shalabh@csa.iisc.ernet.in

Abstract. We propose four variants of a recently proposed multi-timescale algorithm in [1] for ant colony optimization and study their application on a multi-stage shortest path problem. We study the performance of the various algorithms in this framework. We observe that one of the variants consistently outperforms the algorithm in [1].

Keywords: Ant colony optimization, stochastic approximation, multistage shortest path problem.

1   Introduction

Over the last two decades, ant colony optimization (ACO) has emerged as a leading metaheuristic method for the solution of combinatorial optimization (CO) problems. The ACO metaheuristic has been successfully applied in various application domains such as traveling salesman problem (TSP), sequential ordering problem, the quadratic assignment problem etc., see [2]. A detailed survey of the theoretical aspects of ACO can be found in [3]. Certain convergence proofs of ACO algorithms can be found in [4], [5] and [6]. It is shown that ACO algorithms in general suffer from first order deception in the same way as genetic algorithms. Search techniques and evolutionary algorithms applied to CO problems can sometimes be guided away from potentially good zones of the search space. Such harmful bias can be introduced by the choice of problem representation, choice of operators and by the algorithm itself, see [7].

In [8], the relationship between ACO and stochastic gradient descent is studied. It is shown that some empirical ACO algorithms approximate stochastic gradient descent in the space of pheromones. A stable implementation of gradient based reinforcement learning that belongs to the framework of ACO algorithms is proposed there. In [9], an algorithm for routing in mobile ad hoc networks is presented.

In [1], an ACO algorithm in the framework of multi-timescale stochastic approximation is presented and its convergence analysed in detail using the ordinary differential equation (ODE) approach. The algorithm in [1] has two components – a pheromone update scheme and an agent learning scheme. These two components are viewed as coupled stochastic approximation recursions.

* Supported through Grant No.SR/S3/EECE/011/2007 from Department of Science and Technology, Government of India.

E. Altman and A. Chaintreau (Eds.): NET-COOP 2008, LNCS 5425, pp. 37–44, 2009.

We present in this paper four variants of the algorithm proposed in [1] and study the performance of all algorithms on the multi-stage shortest path problem. In two of the variants, we adapt the Q-learning algorithm from the reinforcement learning literature to the ACO framework and study their performance as well. One of the variants that we propose shows the best performance overall and performs consistently better than the MAF-ACO algorithm of [1].

The rest of the paper is organized as follows: Section 2 describes the MAF-ACO model and algorithm of [1] that we implement for performance comparisons with our other algorithms. Section 3 describes the multi-stage shortest path problem on which the algorithms are applied. Section 4 presents our proposed algorithms. In Section 5, we briefly sketch the convergence of the proposed schemes. Section 6 describes the experimental results on the multi-stage shortest path problem. Finally, Section 7 presents the concluding remarks.

2   The Multi-agent Foraging – ACO Algorithm [1]

The foraging model is inspired by the foraging phenomenon observed in nature wherein a colony of ants cooperatively discovers a shortest path between the nest and the food source. Consider a two-node network containing a source node (S) and a destination node (D) and with $d \geq 2$ paths between the two nodes. Let L_i denote the length of the edge i and we assume that each L_i is a positive integer. Each edge i is assumed to consist of m_i stages of unit length (e.g., if $L_i = 10$, then it is a concatenation of 10 stages of unit length and each $m_i{=}10$). Each ant traverses one stage at a time. Associated with each edge i are two parameters – the pheromone trail parameter T_i and an agent learning parameter X_i. The agent learning parameter is suitably normalized to obtain the probability of selection of that path by an ant. A theoretical justification for the presence of the agent learning parameter is given in [1].

Let M denote the total number of artificial ants in the system. Initially at time $t{=}0$, all ants are at node S. Whenever an ant, k, is at node S at time t, it probabilistically chooses one of the edges i to traverse at time $t{+}1$. Each ant traverses one stage at a time. At every stage, ant k decides to wait with probability p or move ahead with probability $1 - p$, for some small p. Thus, a random delay is introduced in the traversal of each ant. Once an ant, k, reaches the destination node, D, by traversing through edge i, the pheromone value T_i associated with edge i is updated by a constant amount. The ant also instantaneously returns to the source node, S, and is ready to begin a new trip.

The overall algorithm [1] consists of two coupled recursions – a pheromone update scheme and an agent learning scheme. Let $a(t)$, $t \geq 0$ be decreasing step-sizes that satisfy

$$\sum_t a(t) = \infty \text{ and } \sum_t a(t)^2 < \infty. \tag{1}$$

Let $\rho \in (0,1)$ be a constant pheromone decay rate. We denote by $T_i(t)$ and $X_i(t)$, the pheromone strength parameter and the agent learning parameter, respectively, at time t on path i, $i = 1, ..., d$. Let Q be the constant quantum of pheromone deposited by an agent once it has traversed any path. Also, let $R_i(t)$ be the number of agents who have finished traversing path i at time t. The algorithm of [1] is as follows: For $i = 1, \ldots, d$, we have

$$T_i(t + 1) = (1 - \rho)T_i(t) + \rho Q R_i(t) \tag{2}$$

$$X_i(t + 1) = X_i(t) + a(t)X_i(t)T_i(t + 1). \tag{3}$$

In (2), $(1 - \rho)T_i(t)$ accounts for the evaporation of pheromone previously deposited along path i, while $\rho Q R_i(t)$ reinforces the pheromone strength when $R_i(t)$ agents pass through path i at time t. $X_i(t)$ $(t \geq 0)$ in (3) denotes the utility associated with path i as learned by the agents after t time steps. The learning parameter update is then suitably normalized to obtain the probability of path selection by ants. Thus, the probability $P_i(t)$ of an ant selecting path i at time t is given by

$$P_i(t) = \frac{X_i(t)}{\sum_{j=1}^{d} X_j(t)}. \tag{4}$$

Thus a path with a higher utility is selected with a higher probability. Due to the decreasing step sizes $a(t)$, the agent learning scheme runs on a slower time scale as compared to the pheromone update scheme. The step-size schedule also decides the trade off between the exploration and exploitation.

Next, we describe how the MAF-ACO algorithm is applied to the foraging model. Initially all pheromone values are set to zero, i.e., $T_i(0) = 0 \ \forall$ i. The initial values of the agent learning parameters are identically set equal to a small positive constant, i.e., $X_i(0) = c$, $\forall$ i. Given a sufficient number of agents, initially at least one agent goes out on every possible path. Each agent after completion of its journey on a path from source to destination, immediately begins a fresh journey on a new path chosen according to the current values of the path selection probabilities given by (4). Thus the same fixed number of M agents circulate in the system.

At some discrete time instant $t > 0$, starting from $t = 0$, with a very high probability, the agents who traversed the shortest path complete the trip before the agents who chose other longer paths. After completion of the trip they deposit a quantum of pheromone on the corresponding path whose index is say k. Thus $T_k(t) > 0$ while $T_j(t) = 0$ for $j \neq k$. This implies that $X_k(t)$ gets a positive increment while all other $X_j(t)$, $j \neq k$, values remain the same. As a result, the probability $P_k(t)$, of agents choosing path k at time $t + 1$ is greater than the probabilities on the other paths. It is shown in [1] that initial bias builds up towards the shortest path k, that in turn drives the algorithm to converge to it.

3 Multi-stage Shortest Path Problem

Consider the graph representation of a generic multi-stage shortest path problem with K stages. Each stage k is denoted by S_k. Stage S_0 contains a single source node S and stage S_{K+1} has a single destination node D. For simplicity, we assume that each node in stage S_k is connected to every node in stage S_{k+1} for $0 \leq k \leq K - 1$ by a unique edge. There is a unique edge (S,i) between the source node S and every node $i \in S_1$. Similarly there is a unique edge (j,D) between every node $j \in S_K$ and the destination node D. Let L_{ij} denote the length of the edge (i,j). The objective is to find the shortest path between S and D.

At any discrete time instant t, for any $0 \leq k \leq K$, $i \in S_k$ and $j \in S_{k+1}$, let $T_{ij}^k(t)$ denote the pheromone trail associated with edge (i,j) and let $X_{ij}^k(t)$ denote the agent learning parameter associated with edge (i,j). The probability with which any ant chooses to move from node i to node j at time t is given by:

$$P_{ij}^k(t) = \frac{X_{ij}^k(t)}{\sum_{l \in N_i} X_{il}^k(t)} \text{ if } j \in N_i, \tag{5}$$

and is 0 otherwise. Here N_i denotes the set of successor states to state i, i.e., the set of nodes j such that there exists an edge (i, j).

The pheromone update scheme and agent learning scheme for the MAF-ACO applied to the multi-stage shortest path problem are as follows.

$$T_{ij}^k(t + 1) = (1 - \rho)T_{ij}^k(t) + \rho Q R_{ij}^k(t) \tag{6}$$

$$X_{ij}^k(t + 1) = X_{ij}^k(t) + a(t)X_{ij}^k(t)T_{ij}^k(t + 1) \tag{7}$$

In (6), $R_{ij}^k(t)$ denotes the number of ants who have completed a trip between S and D at time t and have traversed the edge (i,j) in stage k.

4   Our Algorithms

We present here our variants of the ACO algorithm in [1] and subsequently investigate, via simulation, their performance in relation to the above algorithm. We observe that a slight variation made to the pheromone update scheme in the MAF-ACO algorithm improves its overall performance. We now propose variants to the algorithm (6)-(7) of [1].

Variant A: The pheromone update scheme specified in (6) is used as is. However, the agent learning scheme is now modified to

$$X_{ij}^k(t + 1) = X_{ij}^k(t)(1 + a(t)\frac{T_{ij}^k(t + 1)}{\Sigma_{(i,l) \in N_i} T_{il}^k(t + 1)}). \tag{8}$$

The idea here is to use normalized pheromone strengths in the learning update instead of the pheromone strengths themselves. The equation has been modified to include the weight contributed by the $(i, j)^{th}$ edge as compared to that of other outgoing edges $(i, l) \in N_i$.

Variant B: The Pheromone update scheme is modified here as

$$T_{ij}^k(t + 1) = (1 - \rho)T_{ij}^k(t) + \frac{\rho Q R_{ij}(t)}{L_{ij}}. \tag{9}$$

The agent learning scheme specified in (7) is retained here. In (9), we incorporate the fact that the strength of the pheromone deposited by ants along a link must depend on its length in an inverse fashion. Thus, larger the length of a link, lower would be the pheromone strength along the link and vice-versa. As we observe in our experiments, this variant performs the best amongst all variants that we study, and performs better than the algorithm in [1].

Variant C: The Pheromone update scheme is modified as in (9). We however use the self-adaptive Q-routing algorithm [10] for parameter learning. The Q-routing algorithm is primarily the Q-learning algorithm, applied for finding the shortest path in routing applications, that is popular in the reinforcement learning literature. We associate with each node a learning parameter instead of the agent learning parameter for each edge. Let $Q_{id}(t)$ denote the learning parameter that a node i estimates for the ant to reach the destination D at time t. Each ant at each node takes a decision based on the learning parameter Q associated with its neighbor nodes. In particular, the probability of choosing node j for an ant at node i is given by

$$P_{ij}(t) = \frac{Q_{jd}(t)}{\Sigma_{(i,l) \in N_i} Q_{ld}(t)}. \tag{10}$$

Once the next node is decided based on the above probability, the value of the learning parameter Q associated with node i is updated as per the standard Q-learning algorithm (below)

$$Q_{id}(t+1) = Q_{id}(t) + \eta(T_{ij}(t) + Q_{jd}(t) - Q_{id}(t)), \tag{11}$$

where η is a "learning rate" parameter and $T_{ij}(t)$ denotes the pheromone strength associated with edge (i,j) at time t. The idea here is to use Q-values $Q_{ij}(t)$ to update the probabilities of choosing paths instead of the learning rate parameter $X_{ij}(t)$. Thus a path with a higher Q-value has a higher probability of selection.

Variant D: Both the pheromone update and the agent learning schemes are as in Variant C here. The probabilities $P_{ij}(t)$, however, are selected according to

$$P_{ij}(t) = \frac{T_{ij}(t) + Q_{jd}(t)}{\Sigma_{(i,l) \in N_i}(T_{il}(t) + Q_{ld}(t))}, \tag{12}$$

in place of (10). Note that $(T_{ij}(t) + Q_{jd}(t))$ is an estimate of the value of the edge (i, j) chosen at node i at time t. Thus, while in Variant C, we use the Q-value of the 'next' state as the learning parameter for obtaining the path selection probabilities, in Variant D, we use the value of the 'current' state by summing the single-stage reward (note that the pheromone update now serves as the single-stage reward) and the Q-value of the 'next' state, for the same. Thus, in Variant D, the edge with the maximum value has the highest probability of being picked.

5   A Sketch of Convergence

The analysis follows along the lines of [1]. Hence, only a brief sketch is presented. We consider Variant B first. From (5), upon simplification, one obtains the following recursion for $P_{ij}^k(n)$:

$$P_{ij}^k(n+1) = P_{ij}^k(n) + a(n)P_{ij}^k(n)(T_{ij}^k(n+1) - \sum_{l \in N_i} P_{il}^k(n)T_{il}^k(n+1))(1 - O(a(n))).$$
(13)

The above can be seen to be an Euler discretization (see [1]) of the following ODE:

$$\dot{P}_{ij}^k = P_{ij}^k(f_{ij}^k(P) - \sum_{l \in N_i} f_{il}^k(P)P_{il}^k),$$
(14)

where $f_{ij}^k(P) = E_s[T_{ij}^k(n)]$ is the stationary expectation for the pheromone values. For the iterates as in (9), one can obtain in a similar manner as Lemma 2 of [1] that

$$f_{ij}^k(P) = \frac{P_{ij}^k QM}{L_{ij} \sum_{l \in N_i}(m_l + 1)P_{il}^k}.$$

Let S_{N_i} denote the N_i-dimensional simplex

$$S_{N_i} = \{q_i = (q_{i1}, \ldots, q_{i|N_i|}) \in \mathcal{R}^{N_i} \mid q_{il} \geq 0 \; \forall l \in N_i \text{ and } \sum_{l \in N_i} q_{il} = 1\}.$$

Functions $f_{ij}^k(\cdot)$ are Lipschitz continuous on S_{N_i}, hence (14) is well posed. One can now show as in Lemma 4 of [1] that $(1, 0, \ldots, 0)$ is an asymptotically stable equilibrium for the ODE (14) with domain of attraction $\mathcal{D} = \{x \in S_{N_i} \mid x_1 > x_j, j \neq 1\}$. A similar result as Theorem 1 of [1] then shows that once iteration (13) enters $\mathcal{D}$, it asymptotically converges to the above point with a probability that increases asymptotically (with the iteration index) to one. The convergence for Variant A can be shown in a similar manner.

For Variants C and D, note that boundedness of iterates for the Q-learning recursion is not guaranteed. For purposes of showing convergence of the Q-learning recursion, one can reset iterates as soon as they go outside a unit ball. As we see in our experiments, we do not require the modified procedure suggested above. Since we work only with probabilities (in order to compute the optimal paths), the corresponding recursion for the probabilities remains bounded and one can indeed show that using both – Variants C and D, $P_{ij}(t)$ will converge to one for the edge (i, j) that has the maximum Q-value.

6   Experimental Results

In this section, we report the simulation results using our algorithms and the MAF-ACO algorithm of [1], and compare their performance. The multi-stage shortest path model on which the various algorithms have been applied is shown in Fig. 1. In our simulations, we ran each algorithm for a sufficiently long time and the shortest path found in each execution of each algorithm was recorded. We treat each execution of each algorithm as a *trial*. 100 independent *trials* of each algorithm are run and the number of times the algorithm converged to the shortest path was observed. The performance has been summarized for $M \in \{32, 64, 128, 256\}$ ants. The agent learning parameter and pheromone value associated to each edge are set initially to 1 and 0 respectively. The learning parameter η is set to 0.5. The quantity of pheromone Q is initially set to 1.

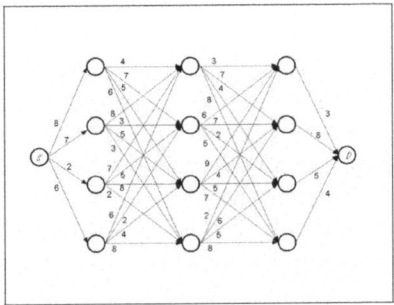

Fig. 1. The directed graph for the multi-stage shortest path problem simulation

The value of ρ was not found to be critical in the experiments and hence was set to 0.1 for all the trials (Table 2). We however also present results (Table 1) of experiments using the original MAF-ACO algorithm and our Variant B for different values of ρ. In the agent learning scheme, the initial value of the step-size was set to $a(0) = 0.9$ and it was updated every 10 iterations by applying the rule: $a(10n) = \frac{a(0)}{n}$ and $a(10n + k) = a(10n)$ for $1 \leq k \leq 10$. This can be seen to satisfy the step-size requirement (1). The choice of ρ plays a significant role in the performance of traditional ACO algorithms. Our algorithms and the MAF-ACO algorithm, however, are quite tolerant to variations in the value of ρ. This can be attributed to the presence of the agent learning scheme which introduces an extra averaging step. It has been observed that in most cases with all the algorithms, where the solution did not converge to the optimal solution, it has converged to the next best solution.

Table 1. The % of correct solutions found by MAF-ACO and Variant B for various values of evaporation rate (ρ) and number of agents (M)

ρ	MAF-ACO				Variant B			
	Number of agents M				Number of agents M			
	32	64	128	256	32	64	128	256
0.1	55	80	95	100	65	88	99	100
0.5	55	88	99	100	67	93	100	100
0.7	57	90	99	100	70	95	100	100

Table 2. The % of correct solutions found by all algorithms for $\rho = 0.1$

	Number of agents M			
	32	64	128	256
MAF-ACO	55	80	95	100
Variant A	53	80	97	100
Variant B	65	88	99	100
Variant C	43	75	89	100
Variant D	56	82	95	100

7   Conclusions

We developed four variants of the recently developed MAF-ACO algorithm and studied their performance on a multi-stage shortest path problem. In three of our variants, we take into account the fact that the increment on the strength of the pheromone update depends inversely as the length of the path. Our Variant B that combines the above with the same learning parameter update as the original MAF-ACO algorithm of [1] is seen to perform the best overall. As future work, we shall study routing applications by considering a network of queues and study the effect of congestion there. We shall also develop ACO algorithms for constrained optimization under inequality constraints involving quality of service (QoS) guarantees.

References

1. Borkar, V.S., Das, D.: A novel ACO algorithm for optimization via reinforcement and initial bias. Swarm Intelligence, special issue on Ant Colony Optimization (to appear, 2008)
2. Dorigo, M., Stutzle, T.: Ant Colony Optimization. MIT Press, Cambridge (2004)
3. Dorigo, M., Blum, C.: Ant colony optimization theory: A survey. TCS: Theoretical Computer Science 345 (2005)
4. Gutjahr, W.J.: A graph-based ant system and its convergence. Future Generation Comp. Syst. 16(8), 873–888 (2000)
5. Stutzle, T., Dorigo, M.: A short convergence proof for a class of ant colony optimization algorithms. IEEE-EC 6, 358–365 (2002)
6. Merkle, D., Middendorf, M.: Modeling the dynamics of ant colony optimization. Evolutionary Computation 10(3), 235–262 (2002)
7. Blum, C., Dorigo, M.: Search bias in ant colony optimization: on the role of competition-balanced systems. IEEE Trans. Evolutionary Computation 9(2), 159–174 (2005)
8. Meuleau, N., Dorigo, M.: Ant colony optimization and stochastic gradient descent. Artificial Life 8(2), 103–121 (2002)
9. Di Caro, G.A., Ducatelle, F., Gambardella, L.M.: AntHocNet: An ant-based hybrid routing algorithm for mobile ad hoc networks. In: Yao, X., Burke, E.K., Lozano, J.A., Smith, J., Merelo-Guervós, J.J., Bullinaria, J.A., Rowe, J.E., Tiňo, P., Kabán, A., Schwefel, H.-P. (eds.) PPSN 2004. LNCS, vol. 3242, pp. 461–470. Springer, Heidelberg (2004)
10. Ghassemlooy, Z., Tekiner, F., Srikanth, T.R.: Comparison of the q-routing and shortest path routing algorithms. In: Proc. of the 5th Annual Postgraduate Symp. on the Convergence of Telecommunications, Networking and Broadcasting (2004)

MPLS Online Routing Optimization Using Prediction

Abutaleb Abdelmohdi Turky and Andreas Mitschele-Thiel

Integrated HW/SW Systems Group, Ilmenau University of Technology
98693 Ilmenau, Germany
{abutaleb-abdelmohdi.turky,mitsch}@tu-ilmenau.de

Abstract. This paper presents an efficient enhancement to the online routing algorithms for the computation of Labeled Switching Paths (LSPs) in Multiprotocol Label Switching (MPLS) based networks. To achieve that, an adaptive predictor is used to predict the future link loads. Then the predicted values are incorporated in the link weights formula. Our contribution is to propose a new idea that depends on the knowledge of the future link loads to achieve a routing that can be done much more efficiently. According to the non-linear nature of traffic, we use a Feed Forward Neural Network (FFNN) to build an accurate traffic predictor that is able to capture the actual traffic behaviour. We study two performance parameters: the rejection ratio and the percentage of accepted bandwidth in different load conditions. Our proposed algorithm in general, reduces the rejection ratio of requests and achieves higher throughput when compared to CSPF and WSP algorithms.

Keywords: MPLS, traffic engineering, routing, neural network.

1 Introduction

The growth of the Internet requires complexity level of network management techniques such as Traffic Engineering (TE) in order to provide the demanded quality of service (QoS). Traffic engineering targets to optimize the performance goals such as delay minimization, throughput maximization and it aims to optimize the resource utilization. The extensions of routing protocols are proposed to modify these protocols for traffic engineering includes connectionless routing protocols such as Open Shortest Path First (OSPF) and the more recently deployed MPLS protocol.

In MPLS networks [1], the packets are assigned with labels at the ingress router and these labels are used to forward the packets according to specific LSPs. Service providers can use LSPs to implement Virtual Private Networks (VPNs) or to satisfy other QoS agreements. There is another signalling protocol such as Recourse Reservation Protocol (RSVP-TE) [2] or Label Distribution Protocol [3] (LDP) that used to setup the paths. In other words, the route of LSPs is explicitly defined between the ingress and egress routers. With the explicit routing of LSPs in MPLS technology, the service providers have an important feature to be able to engineer how their traffic will be routed, and have the ability to improve the network utilization, by minimizing the LSPs rejection ratio when the network is overloaded.

The efficiency of TE schemes mainly depends on route optimization. Routing algorithms [4] can be classified to static or dynamic according to the nature of

E. Altman and A. Chaintreau (Eds.): NET-COOP 2008, LNCS 5425, pp. 45–52, 2009.

information used for selecting LSP routes. Static algorithms depend on fixed information which does not change with time; dynamic algorithms use the current state of the network. Also, routing algorithms can be executed either online or offline according to when the computation is done. In online routing algorithms path requests are attended to one by one request. This paper is focused on dynamic online routing.

In this paper, we introduce a new idea to enhance the routing performance by using FFNN to build an accurate predictor that is able to predict the actual traffic behaviour. Then we use it to predict the future load in every network links and incorporate this value in link weights formula. The remainder of this paper is organized as follows: Section 2 overviews the related work. Section 3 demonstrates the predicting of future load-based routing algorithm. Section 4 discusses the performance evaluation. The future work and our conclusions are presented in section 5.

2 Related Works

The Shortest Path First (SPF) algorithm is the most commonly used algorithm within MPLS Domains. In SPF, the path that contains least number of links between the source and destination pair is selected. Although SPF is trying to minimize resource occupation, it leads to congestions in some network links and causes unbalancing in resource utilization. R. A. Guerin [5] introduced a modification to the shortest path algorithm, called Widest Shortest Path (WSP), which is based on the computation of the shortest paths in the first stage and if there is more than one of those shortest paths, it chooses the one with maximum bandwidth.

E. Crawley [6] proposed the Constraint Shortest Path First (CSPF) protocol to overcome the problem of load balancing in OSPF which modifies the link cost to reflect the current resource availability. The cost of links is inversely proportional to the residual link capacities. Since all previous algorithms choose the LSPs without considering the future LSPs requests, the related performance dose not achieve the best result in maximizing network utilization or request acceptance rate.

An example of advanced routing algorithm is Minimum Interference Routing Algorithm (MIRA) [7]. The idea of MIRA is to prevent routing over links that may interfere with another path requests in the future. The algorithm picks up the minimum interference path for a given LSP that maximizes the minimum max-flow between all other ingress-egress pairs. The computations of MIRA routing are taking long time and MIRA does not work as expected for some network scenarios.

R. Boutaba [8] introduced a dynamic online routing algorithm (DORA) that compute the path potential value (PPV) array associated with a source-destination pair in the first step. PPV is used to avoid routing over links that have high potential to be part of any other paths. Then the algorithm incorporates PPV value with residual link bandwidth to form a weight value for each link that is used to compute a weight-optimized network path. DORA offers better performance than MIRA and WSP algorithms. Additionally computations used in DORA are less expensive.

E. Einhorn and A. Mitschele-Thiel [9] introduced Reinforcement Learning for Traffic-Engineering (RLTE) algorithm. This work presents a novel distributed and self-organized QoS routing approach that is based on reinforcement learning.

3   Predicting of Future Load-Based Routing Algorithm

This section provides a detailed description of Predicting of Future Load-based Routing algorithm (PFLR).Firstly; we want to describe the idea behind design of PFLR. We believe that in any routing algorithm which depend on residual bandwidth in network links in order to choose LSPs: consideration of the future load value will optimize the routing performance more effectively. Therefore, we propose to build an accurate traffic predictor that will be able to predict the actual traffic behaviour.

The future links load depends on many parameters such as network topology, network load condition (inter-arrival rate and holding time of the requests in the network) and the behaviour of selected routing algorithm. Artificial Neural network (ANN) offers prediction capability with different types of network traffic and has ability to learn on line to be adaptive. With experimental results [10], ANN approach can accurately predict a complicated network traffic pattern efficiently.

The operation of PFLR is divided into two stages. The first stage is the prediction of the future available BW on every link after specific period of time WS (Window Size). The operation of predictions should be made decentralized for example; a specific node is responsible for the prediction operations in specific part of the network to acquire fast prediction and to distribute the complexity of prediction.

In the second stage, the predicted value of future load and current residual bandwidth of each link are combined together in specific formula to represent the link weight. The formula is controlled by a parameter called α which represent the prediction weight. Finally, we run the already selected algorithm to compute a weight-optimized path without any change in its behaviour .In other words, we do not change the routing algorithm that is already running in the network, we just modify the links weights while taking in our consideration the future load. Thus our proposed algorithm has the capability to run with any routing algorithm depend on residual bandwidth in its computations such as WSP, CSFP and DORA.

3.1   PFLR Algorithm

Input:
    The network topology and all residual link capacities.
    The LSPs requests between the different ingress-egress pairs.
Output:
    Routed paths through the network.
Algorithm:
   1.   Predict the future available bandwidth in all links in the network after a specified period of time WS.
   2.   Repeat the following steps until the time of WS has elapsed.
       2.1 Construct the link weight using the following equation:

$$W=(1-\alpha)*(1/\text{current available BW})+\alpha*(1/\text{predicted available BW}) \qquad (1)$$

       Where α = (prediction weight)
       2.2 Complete the normal routing algorithm without changing anything.
   3.   Repeat the previous steps and so on.

---

4 Performance Evaluation

In this section, we will demonstrate the set of experiments used to evaluate the performance of PFLR and will discuss the results. All experiment scenarios are implemented using Visual Basic and Neural Network toolbox in MATLAB [11]. We will test PFLR performance and compare it with WSP and CSPF algorithm. Two parameters performance are studied: first is the rejection ratio of LSPs requests and second is the percentage of accepted BW in two different network load conditions.

Also we want to prove our earlier stated hypothesis of consideration of the future load in routing decision in order to enhance the routing performance. Therefore in every scenario we shall compare two routing algorithms with two different types of future information. The first type is the actual value of link loads after a specific period of time. In this case, we assume that the information of future values related to arrival time, the source and destination nodes and the required capacity of the next LSPs are already known. In other words, we shall run the selected algorithm for routing the next requests in the WS period. Then we take the value of link loads at the end of the WS period (as future info.). Then we return back again to the start of WS period in order to combine this value with the current BW value. The second type is the predicted link load that fully depends on the history of traffic in a particular link and does not know any thing about the future (the real case of network routing).

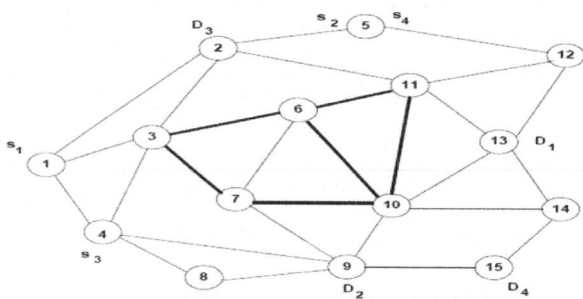

Fig. 1. MIRA network topology

Our experiment is done on MIRA network that is shown in Fig. 1. The thicker links have a capacity of 4800 capacity unit while the thinner links have a capacity of 1200 capacity unit. Fig. 1 shows the location of four different source–destination pairs that are identified by (S1, D1), (S2, D2), (S3, D3), and (S4, D4). The capacities of requested paths are randomly distributed among 10-40 capacity units. We examine the performance for two different network load conditions. The first network load condition is a moderate load condition: when the arrival of requests follows a Poisson distribution with mean $\lambda = 32.5$ requests per time-unit and the holding time of request is based on an Exponential distribution with mean $\mu = 10$ time-units. The second is heavy load condition: when the inter arrival of path requests λ is equal to 35 requests per time-unit and the holding time of request μ is equal to 10 time-units.

The structure of FFNN is shown in Fig. 2. FFNN consists of three layers: The input layer contains 16 neurons that are connected directly to the last 16 traffic sample. The hidden layer consists of 20 neurons and only one neuron in the output which gives the predicted value of link load after WS. The Levenberg-Marquardt training algorithm is used because it is the fastest and most accurate one in our case. FFNN is designed to be adaptive. It takes the previous thousand of traffic samples and train on it to achieve the best predictor design. Then FFNN is used to predict the future link load during the next hundred traffic samples and then it will trains again on the previous thousand of traffic samples and so on.

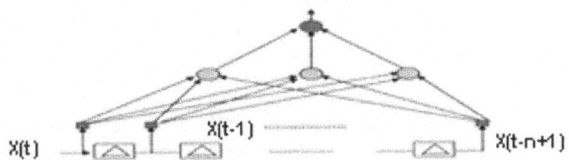

Fig. 2. Feed Forward Neural Network architecture

The optimal values of **α** and WS depend on the behaviour of routing algorithm and the load condition. In every scenario, we search for the optimal values of **α** and WS by trying different values of **α** (range from 0 to 1) and WS (range from 5 to 9). In case of the actual value of link loads, **α** and WS take different values related to every scenario. In case of the predicted value of link loads, we found the fewest rejection ratio at **α**=0.1 and WS=8 in all PFLR scenarios. Table 1 show the rejection ratio of 2000 requests (in CSPF with PFLR) for the moderate load condition scenario only with different combination of **α** and WS.

Table 1. The rejection ratio of requests for different combination of **α** and WS parameters

α	WS=5	W =6	WS=7	WS=8	WS=9
0.0	1,257	1,257	1,257	1,257	1,257
0.1	1,170	1,234	1,277	**1,166**	1,275
0.2	1,328	1,203	1,186	1,198	1,395

4.1 Moderate Load Condition Scenario

Fig. 3 shows the rejection ratio of requests for the moderate load condition. Based on Fig. 3, WSP_FUT (WSP with actual future load at **α**=1.0 and WS=6.) rejects approximately 8.68% less requests than WSP, followed by WSP_PFLR (WSP with Prediction at **α**=0.1 and WS=8.) that rejects approximately 8.51% less requests than WSP. Also CSPF_FUT (at **α**=0.095 and WS=8.) rejects approximately 6.42% requests less than CSPF, followed by CSPF_PFLR (at **α**=0.1 and WS=8.) that rejects approximately 4.45% less requests than CSPF.

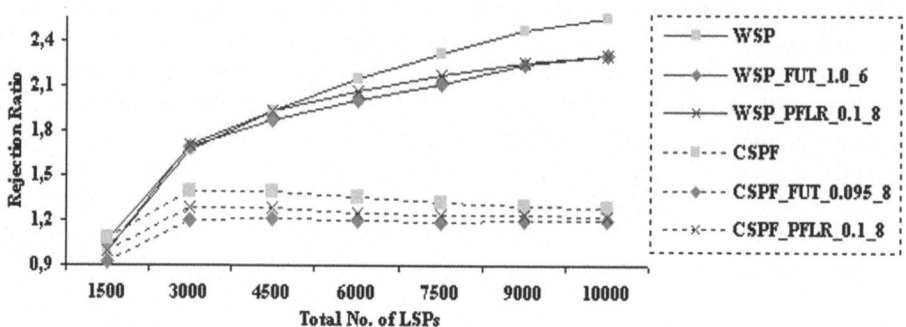

Fig. 3. The rejection ratio of requests for the moderate load condition

Fig. 4 shows the percentage of accepted BW for the moderate load condition. Based on Fig. 4, WSP_PFLR (at $\alpha=0.1$ and WS=8) accepts 0.38% more bandwidth than WSP, followed by WSP_FUT (at $\alpha=1.0$ and WS=6) that accepts 0.35% more bandwidth than WSP. Also CSPF_FUT (at $\alpha=0.095$ and WZ=8.) accepts 0.10% more bandwidth than CSPF, followed by CSPF_PFLR (at $\alpha=0.1$ and WS=8.) accepts 0.06% more bandwidth than CSPF.

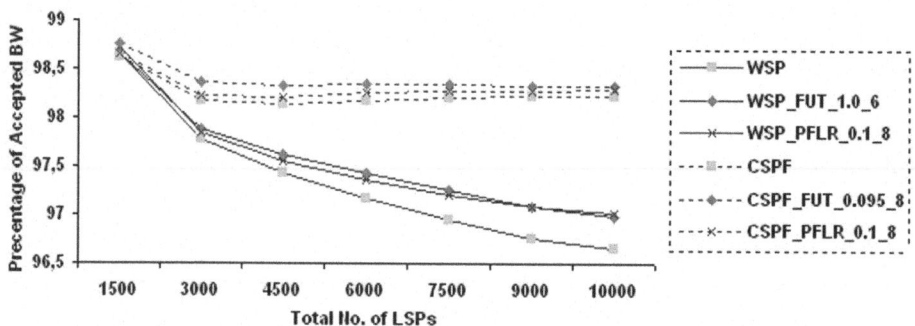

Fig. 4. The percentage of accepted BW for the moderate load condition

4.2 Heavy Load Condition Scenario

Fig. 5 shows the rejection ratio of requests for the heavy load condition. Based on the Fig. 5, WSP_FUT (at $\alpha=1.0$ and WS=6.) rejects approximately 4.45% less requests than WSP, followed by WSP_PFLR (at $\alpha=0.1$ and WS=8.) that rejects approximately 2.58% less requests than WSP. Also CSPF_PFLR (at $\alpha=0.1$ and WS=8) rejects approximately 6.77% less requests than CSPF, followed by CSPF_FUT (at $\alpha=0.205$ and WS=8) that rejects approximately 2.87% less requests than CSPF.

Fig. 6 shows the percentage of accepted BW for the heavy load condition. Based on Fig. 6, WSP_FUT (at $\alpha=1.0$ and WS=8.) accepts 0.29% more bandwidth than WSP, followed by WSP_PFLR (at $\alpha=0.1$ and WS=8.) that accepts 0.13% more bandwidth than WSP. Also CSPF_PFLR (at $\alpha=0.1$ and WZ=8.) accepts 0.57% more

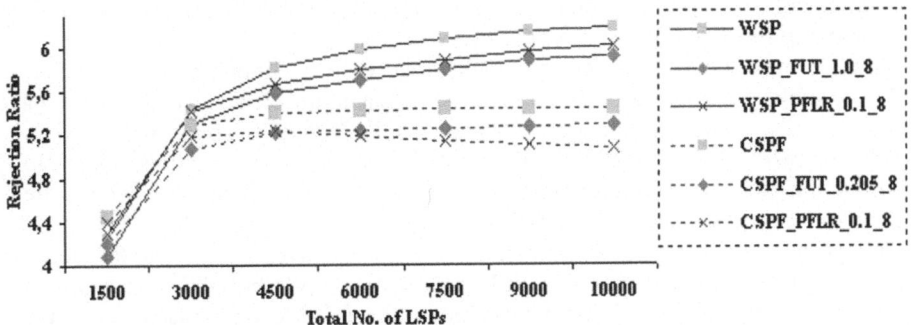

Fig. 5. The rejection ratio of requests for the heavy load condition

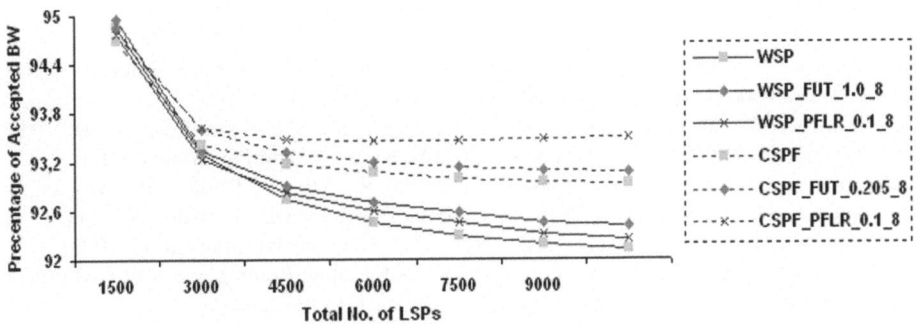

Fig. 6. The percentage of accepted BW for the heavy load condition

bandwidth than CSPF, followed by CSPF_FUT (at $\alpha=0.205$ and WZ=8.) that accepts 0.13% more bandwidth than CSPF.

4.3 Complexity Analysis of PFLR

PFLR algorithm requires additional commotional time to achieve enhanced routing performance. This time consists of two parts, the training time of predictors and the prediction time. As we say before, the predictors are distributed on the nodes. This means every node will responsible for (E/V) operation where E is the number of links and V is the number of nodes. The training operation happens only one time every specific period RP (Retrain Period). The predation also happens every WS period. Thus, the training requires O ((E Tt) / (V RP)) where Tt is the training time of one predictor and the prediction requires O ((E Pt) / (V WS)) where Pt is the prediction time of one predictor. The prediction is computationally inexpensive which is equal to 0.006 sec. But the training requires more time which is equal to 0.078 sec.

5 Conclusion and Future Work

In this paper, we have proposed a new TE algorithm named PFLR that can efficiently enhance the dynamic on line routing algorithms in MPLS-based networks. This algorithm can work with any other online routing algorithms that have computations depending on residual BW in network links.

First, we have proved our theory which suggests using future load information in order to achieve a routing that can be done much more efficiently. Then we have experimentally compared the performance of PFLR modification with WSP and CSPF in two different network load conditions and have shown that PFLR modification performs considerably better than WSP and CSPF algorithms with respect to two performance comparison criteria.

As future work, we plan to test the PFLR performance on a complex network topology. Also we need to test the PFLR performance upon link failure scenario. We need to compare PFLR with other algorithms like DORA too. Finally, we want to focus on decentralized routing algorithms such as any local optimization algorithms.

References

1. Rosen, E., Viswanathan, A., Callon, R.: Multiprotocol Label Switching Architecture. RFC 3031, Network Working Group (2001)
2. Awduche, D., Berger, L., Gan, D., Li, T., Srinivasan, V., Swallow, G.: RSVPTE: Extensions to RSVP for LSP tunnels. RFC 3209, Network Working Group (2001)
3. Jamoussi, B., Andersson, L., Callon, R., Dantu, R., Wu, L., Doolan, P., Worster, T., Feldman, N., Fredette, A., Girish, M., Gray, E., Heinanen, J., Kilty, T., Malis, A.: Constraint based LSP setup using LDP. RFC 3212, Network Working Group (2000)
4. Marzo, J.L., Calle, E., Scoglio, C., Anjah, T.: QoS Online Routing and MPLS Multilevel Protection: A Survey. IEEE Commun. Mag. 41(10), 126–132 (2003)
5. Guerin, R., Orda, A., Williams, D.: QoS routing mechanisms and OSPF extensions. In: IEEE Global Telecommunication, pp. 1903–1908. IEEE Press, Phoenix (1997)
6. Crawley, E., Nair, R., Jajagopalan, B., Sandick, H.: A Framework for QoS-based Routing in the Internet. RFC 2386, Network Working Group (1998)
7. Kar, K., Kodialam, M., Lakshman, T.V.: Minimum Interference Routing of Bandwidth Guaranteed Tunnels with MPLS Traffic Engineering Applications. IEEE J. Selected Areas in Comm. 18(12), 2566–2579 (2000)
8. Boutaba, R., Szeto, W., Iraqi, Y.: DORA: Efficient Routing for MPLS Traffic Engineering. J. Net. and Sys. Man. 10(3), 309–325 (2002)
9. Einhorn, E., Mitschele-Thiel, A.: RLTE: Reinforcement Learning for Traffic-Engineering. In: 2nd International Conference on Autonomous Infrastructure, Man. and Sec., Bremen (2008)
10. Eswaradass, A., Sun, X.H., Wu, M.: Network Bandwidth Predictor (NBP): A System for Online Network performance Forecasting. In: Sixth IEEE International Symposium on Cluster Computing and the Grid, pp. 265–268. IEEE Computer Society, Singapore (2006)
11. Neural Network Toolbox, MATLAP V.7,
    http://www.mathworks.com/products/neuralnet

On the Identifiability of Link Service Curves from End-Host Measurements

Amr Rizk and Markus Fidler*

Multimedia Communications Lab
TU Darmstadt, Germany
{amr.rizk,markus.fidler}@kom.tu-darmstadt.de

Abstract. We estimate service curves of network internal links from end-host measurements of probing traffic. Our approach belongs to the field of network tomography, which deals with the fundamental challenge of identifiability in a priori under-determined network equation systems. As opposed to recent methods that estimate sole quantities, such as delay and bandwidth, we characterize links using the more generic concept of service curve that comprises various derived quantities including the ones mentioned above. Key to our solution is the Legendre-Fenchel transform that achieves additivity of link service curves. Our measurement results reveal that the burstiness of cross traffic flows, which has significant impact on the shape of leftover service curves, can be attributed to individual links. Using the network calculus we show fundamental limits of certain tomography approaches regarding the identification of propagation delays as well as regarding the resolution of post-narrow links.

1 Introduction

Internet service providers largely keep performance measures of their infrastructure confidential. Taking measurements only at end-hosts, network tomography considered here seeks to break results obtained for network paths down to characteristics even of individual links. Per-link metrics including loss, delay, or available bandwidth are vital for numerous applications, such as network monitoring, fault detection, service level agreement verification, and for improving services [3]. For completeness we mention that network tomography can also be applied to synthesize path characteristics from link results, e.g. for origin-destination traffic matrix estimation, a problem which we do not address here.

In this work we seek to identify link service curves from end-host measurements. The notion of service curve and the network calculus provide a general framework for modeling and analysis of transmission systems that goes significantly beyond single performance metrics such as delay or available bandwidth that are primarily used in network tomography today. At the same time service curves facilitate deriving these metrics easily.

---

* This work was funded by the German Research Foundation (DFG) under an Emmy Noether grant.

E. Altman and A. Chaintreau (Eds.): NET-COOP 2008, LNCS 5425, pp. 53–61, 2009.

The network calculus [2,10] is a min-plus system theory for queuing systems that dates back to the seminal works in [4,14]. A system is said to offer a (lower) service curve $S(t)$ to a cumulative traffic arrival flow $A(t)$ if the departures $D(t)$ satisfy $D(t) \geq A \otimes S(t)$ where $\otimes$ denotes the min-plus convolution defined as $A \otimes S(t) = \inf_\tau \{A(\tau) + S(t - \tau)\}$. Here the arrivals $A(t)$ and departures $D(t)$ are the cumulative number of bits entering respectively leaving the system in the time interval $[0, t)$. The service curve $S(t)$ specifies the (minimal) service guarantee over a time interval of duration t. If the system is linear under the min-plus algebra, e.g. a constant rate link or a leaky-bucket traffic regulator, it has an exact service curve $S(t)$ leading to departures given by $D(t) = A \otimes S(t)$. Performance bounds, e.g. for backlog and delay, follow easily from these definitions. Furthermore, whole network paths consisting of tandem systems $S_1, S_2, \ldots S_N$ can be convolved to an end-to-end service curve $S_{e2e}(t) = S_1 \otimes S_2 \otimes \cdots \otimes S_N(t)$. The inverse problem, that is identification of a system's service curve from readings of its arrivals and departures, is solved in [12,13]. Related approaches use max-plus algebra to model and identify discrete event systems, see [11] and references therein.

The remainder of this paper is structured as follows. In Sect. 2 we briefly review network tomography and phrase it in a canonical form. In Sect. 3 we show how the Legendre aka Fenchel transform converts the task of per-link service curve inference into this form. We adapt two particular network tomography methods to service curve identification. We report experimental results in Sect. 4 where we show the benefits of the service curve approach as well as fundamental limits of certain tomography methods. We present brief conclusions in Sect. 5.

2 Related Work on Network Tomography

Given an N-vector of additive link[1] metrics $\mathbf{L}$, e.g. delays, and an $M \times N$-routing matrix $\mathbf{R}$ the M-vector of path metrics $\mathbf{P}$ follows by matrix multiplication as:

$$\mathbf{P} = \mathbf{R} \cdot \mathbf{L} \tag{1}$$

Note that (1) depicts a very general relationship where $\mathbf{P}$ and $\mathbf{L}$ can describe different metrics, such as delay, loss, or even service curves, if the matrix multiplication in (1) is adapted accordingly.

Network tomography is concerned with the inference of link metrics from path measurements. Given[2] $\mathbf{R}$ the task is an inversion problem, i.e. solving (1) for $\mathbf{L}$. For many realistic networks the number of distinct path measurements that can be taken between end-hosts is smaller than the number of links in the network resulting in an under-determined system, i.e. $M < N$ such that (1) can not be simply inverted. Moreover, there exist basic topologies, e.g. a star, where $M \geq N$ but $\mathbf{R}$ is not invertible due to linear dependencies. We discuss two active probing solutions to these challenges. For a broader overview see e.g. [3].

---

[1] We use unidirectional links and do not make any assumptions of symmetry.

[2] Inference of routing can be based on e.g `traceroute` but is not part of this work.

In [16] multicast-based probing methods for per-link delay inference are developed. The correlation of multicast path measurements provides additional information that is used to solve the otherwise under-determined system (1) for **L**. As a simple example consider three links with labels $1, 2, 3$ that span a tree topology. A probe traverses link 1 and is multicast on link 2 and link 3 afterwards. The probe provides information on paths $\langle 1, 2 \rangle$ and $\langle 1, 3 \rangle$ that are correlated owing to link 1. Using probabilistic methods this correlation is key to inferring the characteristics of link 1 and subsequently of links 2 and 3.

While (1) is defined for additive metrics, e.g. delay, it can be easily extended to multiplicative metrics, such as the probability of successful packet transmission respectively of packet loss. Taking the logarithm of a multiplicative metric yields an additive one such that (1) is recovered. It is shown in [1], that the logarithm transform can ease the multicast-based inference of loss, which is indeed a multiplicative metric.

A different technique called tailgating is proposed in [9] to estimate link bandwidths. It is implemented in a tool called `nettimer`. The approach is also used by `STAB` [17] and adopted in related works [7,15]. Tailgating uses probe packets of alternating size. Large "tailgated" packets are each followed by a very small "tailgating" packet. The large packets are set up by the time-to-live or hop-limit field (TTL) to exit the path after a defined link, causing small packets to queue behind them only up to that part of the path. From there on it is assumed that the small packets travel without queuing to the sink. Hence, the small packets carry information on the first part of the path in their time spacing. This is used to infer the bandwidth only of the first part of the path. Given N links in series labeled $1, 2, \ldots, N$ tailgating can approximate $M = N$ path measurements for the paths $\langle 1 \rangle, \langle 1, 2 \rangle, \ldots, \langle 1, 2, \ldots, N \rangle$ so that (1) is invertible.

Related are bandwidth estimation methods, such as `pathchar`, `clink`, `pchar`, `pathneck`, and `bfind`, that use `ICMP` TTL exceeded messages. These are sent back to the source from the point where the TTL expires. Like tailgating some of these methods send large packets to induce congestion that is measured using small packets. However, here the TTL of the small packets and not the large packets is configured to expire after a certain link to create feedback to the source.

Bandwidth is a metric that uses the minimum, i.e. the bandwidth of a path is determined by the narrow link that is the link with the minimal bandwidth. This results in a modification of (1) where addition is replaced by minimum. As opposed to addition there exists, however, no inverse operation for the minimum. This causes fundamental difficulties regarding the inversion of (1) that are reflected in the inability of current techniques to resolve post-narrow links.

Regarding our target of inferring per-link service curves, denoted **L**, from measured path service curves, denoted **P**, we conclude that a transform that maps min-plus convolution to addition (similar to the logarithm transform that maps multiplication to addition) achieves the canonical form (1). This form facilitates application of various tomography algorithms such as the ones described above. In Sect. 3 we show how the Legendre-transform can be effectively used to this end.

3   Inference of Link Service Curves

In this section we develop methods for inference of link service curves from path measurements. We formulate the problem in min-plus algebra and transform the equation system to obtain the canonical form (1) that can be solved by known tomography algorithms. We adapt two inference methods to our task.

The maximum virtual backlog $B_{\max} = \sup_t\{A(t) - D(t)\}$ is the maximum amount of data contained by a system at any time. Note that virtual backlog comprises bits that are queued, e.g. caused by fluctuating burstiness of the cross traffic, as well as bits that are in transmission. For constant rate probes $A(t) = rt$ at a lossless min-plus linear system, i.e. with an exact service curve $S(t)$,

$$B_{\max}(r) = \sup_t \{rt - S(t)\} =: \mathcal{L}_S(r) \tag{2}$$

holds, where $\mathcal{L}_S$ is the Legendre transform of the service curve $S(t)$ [4,5,12,13].

An important property of the Legendre transform is that twofold application $\mathcal{L}\mathcal{L}_S$ returns the convex hull of S, i.e. the largest convex function that is smaller than S. Thus, we generally have $\mathcal{L}\mathcal{L}_S \leq S$ and $\mathcal{L}\mathcal{L}_S = S$ if S is convex.

The properties of the Legendre transform are used in [12,13] to establish a measurement method. Using constant rate probes $A(t) = rt$ at increasing rates r, referred to as a rate scan, the backlog is determined from measurements of $A(t)$ and $D(t)$. Once an estimate $\tilde{B}_{\max}(r)$ is quantified the Legendre transform reveals a service curve estimate $\tilde{S}(t) = \mathcal{L}_{\tilde{B}_{\max}}(t)$ of a min-plus linear system.

Furthermore, the Legendre transform turns the min-plus convolution into a sum, i.e. $\mathcal{L}(S_1 \otimes S_2)(r) = \mathcal{L}_{S_1}(r) + \mathcal{L}_{S_2}(r)$. We use this property to recast the task of per-link service curve identification in standard form according to (1), such that the methods from network tomography can be applied to it.

Using service curves $S(t)$ as a metric for links $\mathbf{L}$ and paths $\mathbf{P}$ gives us an equation system similar to (1) where links are, however, concatenated to paths by min-plus convolution, e.g. $S_{P_1}(t) = S_{L_1} \otimes S_{L_2} \otimes \ldots$. Such systems are difficult to analyze since there exists no inverse operation for the min-plus convolution[3]. Transforming the whole system into the Legendre domain, however, turns all service curves into backlog curves and all min-plus convolutions into sums, e.g. $B_{\max,P_1}(r) = B_{\max,L_1} + B_{\max,L_2} + \ldots$. Hence, we obtain an additive metric that converts the problem into the exact form (1) where $\mathbf{L}$ and $\mathbf{P}$ are vectors of backlog curves for links and paths respectively. Eventually, reapplying the Legendre transform recovers the equivalent service curves.

In summary, we propose a method that comprises the following steps:

1. Perform rate scans between end-hosts to measure per-path backlog curves
2. Infer per-link backlog curves by inversion of (1) that can be under-determined
3. Convert the backlog curves to service curves using the Legendre transform

---

[3] Contradictory to its name, the min-plus de-convolution is **not** an exact inverse of the convolution [12]. Yet, the two operations obey a certain duality [10]. Given an equation system of min-plus convolution statements, we note that by clever expansion of these equations the duality allows deriving **bounds** on the system's solution.

These steps can be conducted using well-known tomography methods, such as multicasting or tailgating, which are detailed in the following subsections.

3.1   Multicast-Based Inference

Multicast-based inference methods, e.g. [1,16], use the correlation of multicast probes to solve otherwise under-determined tomography problems. The algorithm in [16] can infer per-link delay distributions from path delay measurements. Since backlog, like delay, is an additive metric it is straightforward to adapt the algorithm to infer backlog distributions instead.

We estimate per-path backlog distributions for a set of probing rates and execute the algorithm from [16] for each of the rates to infer per-link backlog distributions that are parameterized by the rate. Finally, we determine maximum backlogs and apply the Legendre transform to estimate the service curves.

A precondition for the inference algorithm in [16] is that delay distributions have a non-zero mass at zero. For this reason constant delays, such as propagation delays, cannot be analyzed and measurement data have to be adjusted accordingly. The same limitation applies in our case when inferring backlog distributions where a constant delay, translates to a constant virtual backlog.

For large probing rates that cause overload the backlog of an ideal lossless system becomes infinity. In this case it is not generally possible to apportion all backlogs to individual links. It is conservative to set indeterminable backlogs to an upper bound respectively to infinity. With the order reversing property of the Legendre transform $\mathcal{L}_{\tilde{B}}(t) \leq \mathcal{L}_B(t)$ if $\tilde{B}(r) \geq B(r)$ resulting service curve estimates $\tilde{S}(t) = \mathcal{L}_{\tilde{B}}(t)$ are valid lower service curves, i.e. $\tilde{S}(t) \leq S(t)$.

We simulated the multicast-based inference method from Sect. 3.1 in Matlab using a fluid flow model [18]. The validation produced results of mixed accuracy. The employed inference algorithm from [16] is analytically exact. It assumes, however, that the correlation of multicast probes stems from joint links in the multicast tree only. While the probes are transmitted independently on all other links, it takes a large number of experiments to obtain sufficiently clean data. This effect is also noted in [16]. Compared to single packet probes in [16] we use rate scans that cause a multiple of probing traffic, thus making the method less practicable in our case. Due to space restrictions we omit showing our results.

3.2   Tailgating

Fix some TTL to mark a hop on a given path and denote $S_u = S_1 \otimes \cdots \otimes S_{\text{TTL}}$ and $S_d = S_{\text{TTL}+1} \otimes \cdots \otimes S_N$ the service curve of the sub-path upstream respectively downstream of hop TTL. We will show how close tailgating measurements can approximate S_u and S_d.

Let $T_d = \max\{t \geq 0 : S_d(t) = 0\}$. We decompose $S_d(t) = \delta_{T_d}(t) \otimes S'_d(t)$ with the delay function $\delta_{T_d}(t) = 0$ for $t \leq T_d$ and $\delta_{T_d}(t) = \infty$ otherwise and a service curve $S'_d(t) = S_d(t + T_d)$. Note that $S'_d(t) > 0$ for all $t > 0$ and $S'(0) = 0$.

Denote A_u, A_d the probing arrivals of the sub-path upstream respectively downstream of hop TTL and D_u, D_d the respective departures. If data are not discarded we have $A_d = D_u$ and the end-to-end service curve is $S_{e2e}(t) = S_u \otimes S_d(t)$ so that

$$S_{e2e}(t) = S_u \otimes \delta_{T_d} \otimes S'_d(t). \tag{3}$$

In case of tailgating we have $A_d = D_u/c$ where $c \gg 1$ since the large tailgated packets are discarded after hop TTL. The receiver uses the small tailgating packets to extrapolate the tailgated packets, i.e. it re-scales the departures D_d by c. Assuming exact service curves we have $D_d(t) = A_d \otimes S_d(t)$. It follows by insertion that $cD_d(t) = c(D_u/c \otimes \delta_{T_d} \otimes S'_d(t))$. Substitution of $D_u(t) = A_u \otimes S_u(t)$ and employing the concept of data scaling from [6] we derive $cD_d(t) = A_u \otimes S_u \otimes \delta_{T_d} \otimes cS'_d(t)$. Hence, the tailgating service curve is $S_{tg}(t) = S_u \otimes \delta_{T_d} \otimes cS'_d(t)$ and for large c

$$S_{tg}(t) \xrightarrow[c \to \infty]{} S_u \otimes \delta_{T_d}(t) \tag{4}$$

since $cS'_d(t) \to \delta_0(t)$ where $\delta_0(t)$ is the neutral element of min-plus convolution.

Using tailgated rate scans with a large packet size ratio (in practice $c \approx 20$) we measure $B_{\max,e2e}(r)$ and $B_{\max,tg}(r)$ at the respective TTL. For min-plus linear systems the backlog curves are the Legendre transforms $\mathcal{L}_{S_{e2e}}(r)$ and $\mathcal{L}_{S_{tg}}(r)$. Backwards transform of $\mathcal{L}_{S_{e2e}}(r)$ and $\mathcal{L}_{S_{tg}}(r)$ by reapplying the Legendre transform reveals estimates of $S_{e2e}(t)$ and $S_{tg}(t)$ respectively, while backwards transform of the difference $\mathcal{L}_{S_{e2e}}(r) - \mathcal{L}_{S_{tg}}(r) = \mathcal{L}_{S'_d}(r)$ yields an estimate of $S'_d(t)$. The resulting estimates $S_{tg}(t) = S_u(t - T_d)$ and $S'_d(t) = S_d(t + T_d)$ are time-shifted versions of $S_u(t)$ and $S_d(t)$. Even an ideal tailgating scheme cannot determine the time-shift T_d. This restriction[4] is mirrored by the multicast-based method, see Sect. 3.1.

For certain large rates the end-to-end as well as the tailgating measurement will indicate overload. In this case the difference $\mathcal{L}_{S_{e2e}}(r) - \mathcal{L}_{S_{tg}}(r)$ evaluates to $\infty - \infty$ such that $\mathcal{L}_{S'_d}(r)$ cannot be estimated for these rates. This ambiguity confirms a fundamental limitation regarding the resolution of post-narrow links. As mentioned for the multicast method in Sect. 3.1, indeterminable backlog values can be set to infinity to estimate a lower service curve nevertheless.

4   Validation in Emulab Experiments

We implemented the tailgating method developed in Sect. 3.2 as an extension of the rude/crude packet sender and receiver (rude.sourceforge.net). We validated the method on the Emulab testbed (www.emulab.org) using the topology given in Fig. 1. All links have a capacity of 100 Mbps and a latency of 4 ms. We use two types of cross traffic, constant bit rate (CBR) of 50 and exponential (EXP) of an average rate of 70 Mbps using the D-ITG traffic generator (www.grid.unina.it/software/ITG/) leaving 50 respectively 30 Mbps unused. We report two experiments where A) the CBR cross traffic is mapped to link 2 and the EXP to link 3 and B) EXP to link 2 and CBR to link 3 such that link 3 becomes a post-narrow link. All except the small tailgating packets have a size of 1000 Byte, including 28 Byte UDP/IP header and 18 Byte MAC header.

---

[4] RTTs of TTL exceeded messages provide a known approach for symmetric paths.

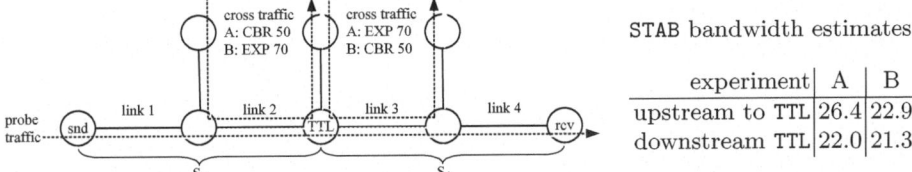

STAB bandwidth estimates

experiment	A	B
upstream to TTL	26.4	22.9
downstream TTL	22.0	21.3

Fig. 1. Topology used for tailgating experiments. The available bandwidths upstream and downstream of TTL are $(50, 30)$ Mbps in experiment A and $(30, 50)$ Mbps in experiment B respectively. The table shows bandwidth estimates from STAB measurements.

Probes, referred to as a rate scan, are CBR packet trains consisting of 400 packets each, sent at rates $5, 10, \ldots, 80$ Mbps. Multiplexing of cross and probing traffic takes place at FIFO schedulers. If not in an overloaded state FIFO systems can be modeled as min-plus linear [13] (a precondition of our method) whereas they become non-linear in case of overload. We use the overload test from [8] to stop the rate scan once overload is detected and set all backlog estimates at subsequent higher rates to infinity.

Fig. 2 shows derivatives of the service curves S_u and S_d upstream and downstream of hop TTL. We display the mean and 0.95 confidence intervals of 100 estimates. Curves labeled "monitor at TTL" and "probe sender at TTL" use a probe monitor respectively sender at hop TTL. These are reference service curves for "tailgating" which assumes no cooperation of hop TTL beyond discarding packets whose TTL expires. Estimates of S_d labeled "inferred" use measurements of the end-to-end path and the path upstream but not downstream of TTL.

Comparing S_u for 50 Mbps CBR cross traffic in Fig. 2(a) with 70 Mbps EXP cross traffic in Fig. 2(c) shows that the long-term rate of S_u converges (within the limits of the probing granularity of 5 Mbps) towards the unused bandwidth of 50 respectively 30 Mbps. Corresponding average available bandwidth estimates of STAB are shown in Fig. 1. Beyond this single value, the service curves show a region of convergence that is dictated by the cross traffic distribution. The burstiness of EXP compared to CBR causes slower convergence where the long-term rate of EXP is achieved only after a relevant time span. Comparing Fig. 2(a) to Fig. 2(b) one can clearly observe the difference in convergence speed that is due to higher burstiness of the cross traffic on link 3 compared to link 2.

Moreover, it can be seen that service is provided only after a certain latency that corresponds to the propagation delay of 4 ms per link. As shown in Sect. 3.2 the tailgating method cannot resolve these delays. Hence, the tailgating estimates of S_u are shifted to the right by $T_d = 8$ ms, which is the delay downstream of hop TTL, see Fig. 2(a) and 2(c). Accordingly, the estimates of S_d are shifted to the left by the same amount, see Fig. 2(b) and 2(d). This restriction is a fundamental limitation of tailgating as shown in Sect. 3.2. Apart from this time-shift the tailgating estimates in Fig. 2(a), 2(b), and 2(c) match the reference service curves accurately. In contrast, the inferred S_d in Fig. 2(d) significantly underestimate the reference estimate generated with a probe sender at hop TTL. This effect reflects a general limitation regarding the resolution of post-narrow links, see also Sect. 3.2.

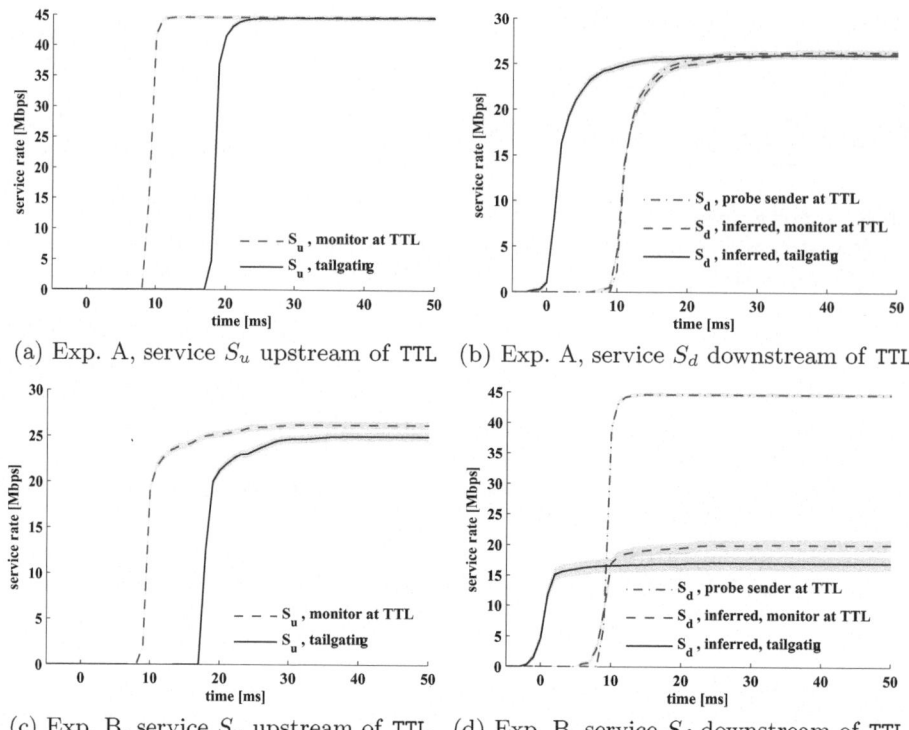

(a) Exp. A, service S_u upstream of TTL   (b) Exp. A, service S_d downstream of TTL

(c) Exp. B, service S_u upstream of TTL   (d) Exp. B, service S_d downstream of TTL

Fig. 2. Per-link service curve estimates. Tailgating can successfully attribute the short-term service convergence caused by the burstiness of cross traffic as well as the long-term available bandwidth to individual links. The measurements confirm the results in Sect. 3.2 that tailgating cannot identify propagation delays and post-narrow links (d).

As an additional reference theory provides a leftover service curve model, see e.g. [10], that can be easily applied to CBR cross-traffic. The theoretical leftover service curve is of the latency rate type $S(t) = R \cdot \max\{0, t-T\}$ where in our case a rate $R = 50$ Mbps and a latency $T = 8$ ms applies. Regarding Fig. 2(a) the service curve estimate with monitor at TTL matches closely, whereas the service curve from tailgating is a time-shifted version as discussed above.

5   Conclusions

We showed how per-link service curves can be inferred from path measurements. Our work builds on a known duality between service curves and backlogs that is established by the Legendre transform. Backlogs are an additive link metric and owing to this property known tomography approaches like multicast probing and tailgating can be effectively adapted for per-link service curve inference. Our measurement results show that estimated service curves provide a higher level of detail compared to available bandwidths where the shape of the service curves reflects the burstiness of cross traffic. The evaluation confirms that our method can attribute these characteristics to individual links.

References

1. Caceres, R., Duffield, N.G., Horowitz, J., Towsley, D.: Multicast-based inference of network-internal loss characteristics. IEEE Trans. Inform. Theory 45(7), 2462–2480 (1999)
2. Chang, C.-S.: Performance Guarantees in Communication Networks. Springer, Heidelberg (2000)
3. Coates, M., Hero, A., Nowak, R., Yu, B.: Internet tomography. IEEE Signal Processing Magazine 19(3), 47–65 (2002)
4. Cruz, R.L.: A calculus for network delay, Part I and Part II. IEEE Trans. Inform. Theory 37(1), 114–141 (1991)
5. Fidler, M., Recker, S.: Conjugate network calculus: A dual approach applying the legendre transform. Computer Networks 50(8), 1026–1039 (2006)
6. Fidler, M., Schmitt, J.B.: On the way to a distributed systems calculus: A network calculus with data scaling. In: ACM SIGMETRICS, June 2006, pp. 287–298 (2006)
7. Harfoush, K., Bestavros, A., Byers, J.: Measuring bottleneck bandwidth of targeted path segments. In: IEEE INFOCOM, April 2003, pp. 2079–2089 (2003)
8. Jain, M., Dovrolis, C.: End-to-end available bandwidth: Measurement methodology, dynamics, and relation with TCP throughput. IEEE/ACM Trans. Networking 11(4), 537–549 (2003)
9. Lai, K., Baker, M.: Measuring link bandwidths using a deterministic model of packet delay. In: ACM SIGCOMM, October 2000, pp. 283–294 (2000)
10. Le Boudec, J.-Y., Thiran, P.: Network Calculus A Theory of Deterministic Queuing Systems for the Internet. Springer, Heidelberg (2001)
11. Lhommeau, M., Hardouin, L., Ferrier, J.-L., Ouerghi, I.: Interval analysis in dioid: Application to robust open-loop control for timed event graphs. In: CDC - ECC 2005, December 2005, pp. 7744–7749 (2005)
12. Liebeherr, J., Fidler, M., Valaee, S.: Min-plus system interpretation of bandwidth estimation. In: Proc. IEEE INFOCOM (May 2007)
13. Liebeherr, J., Fidler, M., Valaee, S.: A system theoretic approach to bandwidth estimation. Technical Report arXiv:0801.0455v1 (January 2008)
14. Parekh, A.K., Gallager, R.G.: A generalized processor sharing approach to flow control in integrated services networks: The single-node case. IEEE/ACM Trans. Networking 1(3), 344–357 (1993)
15. Pasztor, A., Veitch, D.: Active probing using packet quartets. In: ACM IMC, November 2002, pp. 293–305 (2002)
16. Presti, F.L., Duffield, N., Horowitz, J., Towsley, D.: Multicast-based inference of network-internal delay distributions. IEEE/ACM Trans. Networking 10(6), 761–775 (2002)
17. Ribeiro, V., Riedi, R., Baraniuk, R.: Locating available bandwidth bottlenecks. IEEE Internet Computing 8(5), 34–41 (2004)
18. Rizk, A.: Network tomography using min-plus system theory. Master's thesis, TU Darmstadt (March 2008)

Opportunistic Transmission over Randomly Varying Channels

Vivek S. Borkar*

School of Technology and Computer Science,
Tata Institute of Fundamental Research,
Homi Bhabha Road, Mumbai 400005, India
borkar@tifr.res.in

Abstract. This article briefly surveys a connected body work of the author and his collaborators on opportunistic transmission over a randomly varying channel. Both single user and multi-user scenarios are considered and a reinforcement learning based scheme proposed for both, with provable convergence properties.

Keywords: Randomly varying channels, opportunistic transmission, Markov decision processes, structural properties, reinforcement learning.

1 Introduction

Transmission across randomly varying channels calls for opportunistic schemes which transmit more when the channel is favorable and less when it is not, while queuing the packets as needed to facilitate this. Clearly, queuing leads to delay whereas transmission across a channel at a given rate involves power consumption that increases when the channel degrades. This implies a trade-off between power consumption and delay. This problem was first addressed in the single user case in the landmark paper of Berry and Gallager [4]. There have been several subsequent variations and improvements on this work and an extensive literature survey can be found in [14]. My aim here is to summarize a 'learning' approach to this problem that has been developed in [14], [15], which in turn depends on the theoretical groundwork laid in [3], [8]. The aim there is to address the realistic problem of transmission with unknown channel characteristics, on top of the aforementioned issue of the latter varying stochastically with time. When the channel characteristics are known, the standard approach has been to cast the problem as a constrained Markov decision problem [2], which seeks to minimize one cost (such as average power consumption) while imposing a bound on another cost (such as average delay). When the channel characteristics are unknown, one has to build on top of this a learning scheme. We use reinforcement learning based on the notion of 'post-state'. This is proposed, analyzed and

---

* Work supported in parts by a grant from CEFIPRA and a J. C. Bose Fellowship. This is an overview of work done over a period, some of which is joint with Mukul Agarwal, Abhijit Bhorkar, Abhay Karandikar and Nitin Salodkar.

E. Altman and A. Chaintreau (Eds.): NET-COOP 2008, LNCS 5425, pp. 62–69, 2009.

supported by simulations for the single user case in [14] and for the multi-user case in [15]. The latter in turn crucially uses the structural properties of the optimal policies for the constrained Markov decision problem that are established in [3]. Its convergent learning scheme is based on cooperative dynamics as in [8].

The next section describes the single agent problem and its formulation as a Markov decision process. It also summarizes the results of [3] concerning the optimal policy. Section 3 describes the learning scheme for the single user scenario from [14]. Section 4 sketches its extension to the multiuser case from [15]. Section 5 summarizes the conclusions.

2   Single User: Structural Properties

In the single user case, the queue dynamics is given by

$$Q_{n+1} = Q_n - U_n + W_{n+1}, \ n \geq 0, \tag{1}$$

where $\{Q_n\}$ is the queue length at the beginning of n-th slot. $U_n \leq Q_n \wedge R$ is the number of packets transmitted during this slot, R being the maximum number of packets that can be transmitted in a slot, and W_{n+1} is the i.i.d. arrival process with law ζ. We assume that there is a finite buffer with a very large capacity so that buffer overflow effects can be ignored. (This is purely for sake of simplicity.) We assume that the channel state $\{X_n\}$ is a finite state irreducible Markov chain with state space $S \subset (0, \infty)$ and transition probabilities $\kappa(\cdot|\cdot)$. We assume $\{Q_n\}$ to be non-negative integer valued. The aim is to minimize the long run average cost

$$\limsup_{n \uparrow \infty} \frac{1}{n} \sum_{m=0}^{n-1} E[X_m f(U_m)], \tag{2}$$

subject to the constraint

$$\limsup_{n \uparrow \infty} \frac{1}{n} \sum_{m=0}^{n-1} E[Q_m] \leq C, \tag{3}$$

Here $C > 0$ is a prescribed constant and f is a strictly convex strictly increasing function taking positive values such that the power as a function of the channel state x and the number u of packets transmitted is given by $xf(u)$. This expression matches the information theoretic formula for power if one sets $x =$ the inverse norm square of the complex channel gain and $f(u) = c(2^{c'u} - 1)$ for suitable $c, c' > 0$ [15]. We seek to minimize average power consumption subject to a constraint on average queue length, which by a standard argument akin to Little's theorem is equivalent to putting a constraint on average delay. From known results in the theory of constrained Markov decision processes [2], [6], we can consider the equivalent single objective Markov decision process that seeks to minimize

$$\limsup_{n \uparrow \infty} \frac{1}{n} \sum_{m=0}^{n-1} E[X_m f(U_m) + \lambda Q_m], \tag{4}$$

where $\lambda > 0$ is the Lagrange multiplier.

In [3], this problem is considered for more general state spaces for queue length (positive real line) and the channel state (it also considers Markov arrivals), and the dynamic programming equation for the long run average control problem with cost (4) is rigorously justified. See [10] for analogous results for the finite state case. For the discrete case, this equation is

$$V(q, x) = \min_u [c(q, x, \lambda, u) - \beta + \sum_{q', y} p(q', y | q, x, u) V(q', y)], \quad q \geq 0, x \in S, \quad (5)$$

where $p(\cdot)$ is the controlled transition kernel for the controlled Markov chain $\{(Q_n, X_n)\}$, and $c(q, x, \lambda, u) \stackrel{def}{=} x f(u) + \lambda q$. Under suitable irreducibility hypotheses, (5) specifies β uniquely as the optimal cost and V uniquely up to an additive scalar.

Using this and arguments based on convexity and supermodularity, [3] proves that the optimal transmission policy is stationary, i.e., depends only on the current state, and is 'stable', i.e., leads to a positive recurrent Markov chain. (Of course, this requires that we assume the existence of at least one stable stationary policy, which we do.) Furthermore, it has the property that the optimal number of packets to be transmitted in a slot is an increasing function of the queue length and a decreasing function of the channel state.

3   Single User: The Learning Scheme

In principle, the optimal policy can be obtained from (5), which in turn can be solved by the relative value iteration (assuming λ to be known) given by [13]

$$V_{n+1}(q, x) = \min_u [c(q, x, \lambda, u) - V_n(q_0, x_0) + \sum_{q', y} p(q', y | q, x, u) V_n(q', y)], \quad n \geq 0.$$
$$(6)$$

Here (q_0, x_0) is an arbitrarily prescribed state and V_0 is arbitrary.

The problem arises when the statistics of $\{X_n\}$ and / or $\{W_n\}$ is unknown. The reinforcement learning based algorithms for approximate dynamic programming are precisely meant for this situation [5], [12]. In the present case, it is convenient to use the so called post-state framework. We define the post-state as the 'state' immediately after the control action, but before the action of noise, i.e., as $\tilde{Q}_n \stackrel{def}{=} Q_n - U_n$. Then viewing $\{(\tilde{Q}_n, X_n)\}$ as the controlled Markov chain in place of $\{(Q_n, X_n)\}$, the dynamic programming equation becomes

$$\tilde{V}(q, x) = \sum_{w, x'} \zeta(w) \kappa(x' | x) \min_{u \leq q+w} [c(q + w, x', \lambda, u) - \beta + \tilde{V}(q + w - u, x')]. \quad (7)$$

Once again, under suitable irreducibility hypotheses, (7) specifies β uniquely as the optimal cost and $\tilde{V}$ uniquely up to an additive scalar. The advantage here is that the minimization is now inside the averaging operation. The corresponding relative value iteration would be

$$\tilde{V}_{n+1}(q, x) = \sum_{w, x'} \zeta(w) \kappa(x' | x) \min_{u \leq q+w} [c(q + w, x', \lambda, u) - \tilde{V}_n(q_0, x_0)$$
$$+ \tilde{V}_n(q + w - u, x')]. \quad (8)$$

Taking advantage of the preceding observation, one can write a stochastic approximation counterpart of this by first replacing the average on the right hand side by an actual evaluation at an observed transition, and then making an incremental move towards this 'correction', letting stochastic approximation to do the averaging. Specifically, the scheme is

$$V_{n+1}(q, x) = (1 - a(n))V_n(q, x) + a(n)I\{Q_n = q, X_n = x\} \times$$
$$\min_u[c(q + W_{n+1}, x, \lambda, u) + V_n(q + W_{n+1} - u, X_{n+1}) - V_n(q_0, x_0)]$$
$$= V_n(q, x) + a(n)I\{Q_n = q, X_n = x\} \min_u[c(q + W_{n+1}, x, \lambda, u)$$
$$+ V_n(q + W_{n+1} - u, X_{n+1}) - V_n(q_0, x_0) - V_n(q, x)].$$

Here $\{a(n)\}$ is a stepsize sequence satisfying the usual conditions in stochastic approximation: $\sum_n a(n) = \infty$, $\sum_n a(n)^2 < \infty$. We consider the scheme

$$V_{n+1}(q, x) = V_n(q, x) + a(n)I\{Q_n = q, X_n = x\} \min_u[c(q + W_{n+1}, x, \lambda_n, u)$$
$$+ V_n(q + W_{n+1} - u, X_{n+1}) - V_n(q_0, x_0) - V_n(q, x)], \tag{9}$$

with λ no longer constant. In fact, it is replaced by λ_n, which is updated by a separate iteration

$$\lambda_{n+1} = \Gamma\Big(\lambda_n + b(n)(Q_n - C)\Big). \tag{10}$$

Here Γ is a projection to the interval $[0, M]$ for a large M, which serves to keep the iterates bounded. Also, $\{b(n)\}$ satisfies: $\sum_n b(n) = \infty$, $\sum_n b(n)^2 < \infty$, and in addition, $b(n) = o(a(n))$. The latter ensures that (10) moves on a slower timescale than (9) and the 'two time-scale' analysis of [9], Chapter 6, applies.

As described in [9], we can analyze (9) by freezing $\lambda_n \approx$ a constant λ and looking at the limiting o.d.e.

$$\dot{y}(t) = \Lambda(t)(h(y(t)) - y(t)),$$

where $\Lambda(t)$ is a diagonal matrix with nonnegative elements summing to 1 on the diagonal, and $h(y) = [h_{q,x}(y)]$ is given by

$$h_{q,x}(y) = \sum_{w,x'} \zeta(w)\kappa(x'|x) \min_u[c(q + w, x, \lambda, u) + y_{q+w-u,x'} - y_{q_0,x_0}].$$

Under reasonable conditions, the diagonal elements of $\Lambda(t)$ remain uniformly bounded away from zero. One can then prove the convergence of $y(t)$ to the solution $\tilde{V}$ of (7) corresponding to $\tilde{V}(q_0, x_0) = \beta$ along the lines of [1]. In turn, the 'scaled limit' as in [9], Chapter 3, of the above o.d.e. is the same o.d.e. as the above, but with $c(\cdot) \equiv 0$. This has the origin as its globally asymptotically stable equilibrium. Arguing as in [9], Chapters 2 and 3, one concludes that $\sup_n V_n < \infty$ a.s., and furthermore, $V_n \to \tilde{V}$, defined as above, a.s.

However, this analysis treated $\lambda_n \approx$ a constant, so what one really has is that $\{V_n\}$ closely track $\{\tilde{V}_{\lambda_n}\}$, where $\tilde{V}_\lambda$ is $\tilde{V}$ with its λ-dependence made explicit. By an argument adapted from [7] which uses the 'envelope theorem', one can argue that the limiting o.d.e. for the λ_n's is in fact a steepest ascent for the Lagrangian minimized over the primal variables. This converges to the correct Lagrange multiplier, which is the corresponding maximizer. Then so do the λ_n's, a.s. This completes the convergence proof.

4   The Multi-user Scenario

Suppose that there are $K > 0$ users competing for access to a common channel, which is allocated to exactly one of them at a time by a central receiver. Let $\{Q_n(i)\}, \{X_n(i)\}, \{\xi_n(i)\}, \{W_n(i)\}$ denote the corresponding queue length, channel state, departure and arrival processes. Write $Q_n = [Q_n(1), \cdots, Q_n(K)]^T$, with analogous convention for ξ_n's, X_n's and W_n's. Suppose that each user (say, i-th) employs the above scheme to arrive at time n with a quantity $U_n(i) \in [0, Q_n(i) \wedge R]$, which is what she wishes to transmit. She can do so if the channel is allocated to her, but can transmit nothing otherwise. We assume that the channel is allocated to the user with the maximum demand, i.e., to the j-th user where $j = \mathrm{argmax}_i U_n(i)$, at time n. (In case of more than one candidate, one of them may be picked at random.) We wish to analyze this scheme. It is convenient (as in machine learning literature) to work with 'softmax', i.e., the channel is allocated to user i with probability

$$F_i(U_n) \stackrel{def}{=} \frac{U_n(i)^m}{\sum_j U_n(j)^m}, \tag{11}$$

where $m \geq 1$ is a prescribed integer. Thus $\xi_n(i) = U_n(i)$ if the channel is allocated to i and $= 0$ otherwise. In the $m \uparrow \infty$ limit, one gets the original scheme.

We now adapt the arguments of [8] to analyze the behavior of the overall scheme. The analysis of the learning schemes proceeds exactly as before. The fact that the channel is not available at each instant only affects the manner in which the states are being sampled and therefore the process $\Lambda(\cdot)$. As long as $\Lambda(\cdot)$ satisfies the stipulated conditions, viz., that its diagonal elements remain uniformly bounded away from zero, the conclusions are not affected. The queue length dynamics for user i is then

$$Q_{n+1}(i) = Q_n(i) - \xi_n(i) + W_{n+1}(i), \ n \geq 0.$$

Let $\gamma(i) = E[W_n(i)]$. Then the above can be rewritten as

$$Q_{n+1}(i) = Q_n(i) + (\gamma(i) - F_i(U_n)U_n(i)) + \Big((W_{n+1}(i) - \gamma(i))$$
$$+ (F_i(U_n)U_n(i) - \xi_n(i))\Big).$$

Let $\tilde{F} = [\tilde{F}_1, \cdots, \tilde{F}_K]^T$ be defined by: $\tilde{F}_i(x) = F_i(x)x_i \ \forall i$. We assume that this is continuously differentiable. Then the above is of the form

$$Q_{n+1}(i) = Q_n(i) + (\gamma(i) - \tilde{F}_i(U_n)) + M_{n+1}, \ n \geq 0, \tag{12}$$

where $\{M_n\}$ is a martingale difference sequence. Recall that $U_n(i)$ depends on $(Q_n(i), X_n(i))$, i. e., $U_n(i) = \ell_i(Q_n(i), X_n(i))$ for some $\ell_i(\cdot)$. We assume $\ell_i(\cdot, x)$ to be continuously differentiable. Write $\ell(Q_n, X_n)$ for

$$[\ell_1(Q_n(1), X_n(1)), \cdots, \ell_K(Q_n(K), X_n(K))]^T.$$

Let π denote the (unique) stationary distribution of the Markov chain $\{X_n\}$. Define $\hat{F} = [\hat{F}_1, \cdots, \hat{F}_K(\cdot)]^T$ by: for $q = [q(1), \cdots, q(K)]^T$,

$$\hat{F}_i(q) = \sum_x F_i(\ell(q, x))\ell_i(q(i), x(i))\pi(\{x\}).$$

We analyze (12) by looking at its 'fluid' approximation, which is given by the o.d.e.

$$\dot{\tilde{q}}(t) = \gamma - \hat{F}(\tilde{q}(t)), \tag{13}$$

restricted to the positive orthant (i. e., it follows the projected dynamics on the boundary of this orthant). One way to arrive at this is to consider the scaled version of the discrete iteration above, given by

$$Q_{n+1}(i) = Q_n(i) + \eta[(\gamma(i) - \tilde{F}_i(U_n)) + M_{n+1}], \ n \ge 0, \tag{14}$$

for a small $\eta > 0$. The idea is that we consider smaller and smaller time slots of width η with $\gamma(i)$, resp. $\tilde{F}_i$ being 'rates per unit time' rather than 'per slot' quantities. We shall assume that

$$\sum_i \gamma(i) < R, \tag{15}$$

which ensures existence of stable stationary policies (see below). With η small, one may think of (14) as a constant stepsize stochastic approximation iteration analyzed in Chapter 9 of [9]. As argued there, one can conclude that with probability $1 - O(\eta)$, $\{Q_n\}$ will concentrate near the set of equilibria of (13) were the latter itself converging to this set. But our results on structural properties imply that (13) is a cooperative dynamics, i.e., $\frac{\partial \hat{F}_i}{\partial q_j} > 0$ for $i \ne j$. By the Hirsch theorem [11], if its trajectories approach a bounded set, then they converge for almost all initial conditions. Under mild additional assumptions, one can argue as in [9], Chapter 10, that (12) will then concentrate near the set of equilibria with a high ($\approx 1 - O(\eta)$) probability.

This leaves the task of establishing the fact that the trajectories do indeed approach a bounded set. Here's a suggestive, though not completely rigorous argument: Let A_n denote the set of maximizers of $i \to U_n(i)$. Then for $m >> 1$, $F_i(Q_n, X_n) \approx \frac{1}{|A_n|}I\{i \in A_n\}$. Let $\mathcal{V}(q) \stackrel{def}{=} \sum_i q_i$. Thus

$$\dot{\mathcal{V}}(\tilde{q}(t)) = \sum_i \gamma(i) - \sum_i \hat{F}_i(\tilde{q}(t))$$

$$\approx \sum_i \gamma(i) - \sum_x \pi(\{x\})\ell_{i^*(t,x)}(q_{i^*(t,x)}(t), x(i^*(t,x))),$$

where $i^*(t, x)$ is the maximizer of $i \to \ell_i(\tilde{q}_i(t), x(i))$. If any $\tilde{q}_i(t)$ becomes very high, then one expects the r.h.s. $\approx \sum_i \gamma(i) - R$, thus leading to

$$\dot{\mathcal{V}}(\tilde{q}(t)) \leq \sum_i \gamma(i) - R < 0.$$

Thus $\mathcal{V}$ serves as a Liapunov function for (12), leading to bounded trajectories.

To summarize, the scheme will allocate the channel with very high probability to one or more users whose rate requirement is the highest. In [8], the analogy between this and the so-called Wardrop equilibrium in transportation theory has been pointed out. Wardrop equilibrium means that only those routes that have the minimum disutility (i.e., delay / congestion) get traffic. Analogously, here only those users with the maximum rate requirement get the channel. Thus we can think of this as an approximate Wardrop-like equilibrium. Note that when a user with lower rate requirement is not allocated the channel for a while, her queue length and therefore the rate requirement will build up and she will eventually be allocated the channel. The equilibria q^* of the limiting o.d.e. correspond to situations where $\gamma(i) = \hat{F}_i(q^*) \; \forall i$, i.e., the mean arrivals and departures balance out as they should. The discussion above then shows that the stochastic behavior fluctuates around this. Two issues that need further study are an estimate of the actual channel utilization by the users and incentive compatibility of the scheme.

5   Conclusions

This is a brief survey of research done and in progress on the problem of opportunistic communication across a randomly varying channel when the channel characteristics themselves have to be learnt in real time. While the present account confines itself to the theoretical analysis, there are also accompanying simulation results in [14], [15], which give empirical justification for the scheme. They suggest that this line of research is a promising approach to this and related problems.

References

1. Abounadi, J., Bertsekas, D.P., Borkar, V.S.: Learning algorithms for Markov decision processes with average cost. SIAM Journal of Control and Optimization 40(3), 681–698 (2001)
2. Altman, E.: Constrained Markov Decision Processes. Chapman and Hall / CRC, Boca Raton (1999)
3. Agarwal, M., Borkar, V.S., Karandikar, A.: Structural properties of optimal transmission policies over a randomly varying channel. IEEE Transactions on Automatic Control 53(6), 1476–1491 (2008)
4. Berry, R.A., Gallager, R.G.: Communication over fading channels with delay constraints. IEEE Trans. on Information Theory 48(5), 1135–1149 (2002)

5. Bertsekas, D.P., Tsitsiklis, J.N.: Neuro-Dynamic Programming. Athena Scientific, Belmont (1996)
6. Borkar, V.S.: Convex analytic methods in Markov decision processes. In: Shwartz, A., Feinberg, E. (eds.) Handbook of Markov Decision Processes, pp. 347–375. Kluwer Academic, Dordrecht (2002)
7. Borkar, V.S.: An actor-critic algorithm for constrained Markov decision processes. Systems and Control Letters 54(3), 207–213 (2005)
8. Borkar, V.S.: Cooperative dymanics and Wardrop equilibria. Systems and Control Letters 58(2), 91–93 (to appear, 2009)
9. Borkar, V.S.: Stochastic Approximation – A Dynamical Systems Viewpoint. Hindustan Publ. Agency/Cambridge Uni. Press, New Delhi/Cambridge (2008)
10. Djonin, D.V., Krishnamurthy, V.: Structural results on the optimal transmission scheduling policies and costs for correlated sources and channels. In: IEEE Conference on Decision and Control, pp. 3231–3236 (2005)
11. Hirsch, M.W.: Systems of differential equations that are competitive or cooperative II: convergence almost everywhere. SIAM Journal of Mathematical Analysis 16(3), 423–439 (1985)
12. Powell, W.: Approximate Dynamic Programming. Wiley Interscience, Hoboken (2007)
13. Puterman, M.: Markov Decision Processes. John Wiley, New York (1994)
14. Salodkar, N., Bhorkar, A., Karandikar, N., Borkar, V.S.: An on-line learning algorithm for energy efficient delay constrained scheduling over a fading channel. IEEE Trans. on Selected Areas in Communications 26(4), 732–742 (2008)
15. Salodkar, N., Karandikar, A., Borkar, V.S.: A stable on-line algorithm for energy efficient multi-user scheduling (in preparation)

On the Variance of the Least Attained Service Policy and Its Use in Multiple Bottleneck Networks

Matthias Auchmann[1] and Guillaume Urvoy-Keller[2]

[1] Technische Universität Wien, Austria
m.auchmann@artech.at
[2] Eurecom, France
Guillaume.Urvoy@eurecom.fr

Abstract. Size-based scheduling has proved to be effective in a lot of scenarios involving Internet traffic. In this work, we focus on the Least Attained Service Policy, a popular size-based scheduling policy. We tackle two issues that have not received much attention so far. Firstly, the variance of the conditional response time. We prove that the classification proposed by Wierman et al. [11], which classifies LAS as an always unpredictable policy, is overly pessimistic. We illustrate the latter by focusing on the M/M/1/LAS queue. Secondly, we consider LAS queues in tandem. We provide preliminary results concerning the characterization of the output process of an M/M/1/LAS queue and the conditional average response time of LAS queues in tandem.

1 Motivation

Size-based scheduling has proved to be very effective in increasing performance in a lot of scenarios: Web servers [4], Internet traffic [8] or 3G networks [6]. The key idea behind size-based scheduling is to favor short jobs while ensuring that large jobs do not starve. The net result is better interactivity from the user point of view as short jobs correspond to interactive applications, while large job correspond to bulk transfers when considering Internet traffic. The extent to which large jobs suffer depends on the statistical characteristics of the job size distribution and especially onhow the mass is distributed among short and large jobs. Broadly speaking, the larger the mass carried by the large flows, the smaller the penalty since short flows, that have the highest priority, can not monopolize the server. Heavy-tailed distributions, which have often been observed in the Internet [3], feature such a property.

In this paper, we consider the Least Attained Service (LAS) policy, a.k.a the Foreground-Background policy [7]. LAS has been initially proposed and studied in the context of time-sharing computers in the late 60s [10]. Under LAS, priority is given to the job that has received the least amount of service. In case of ties, jobs share the server in a round-robin manner. A salient feature of LAS is that it has no internal parameter to tune.

E. Altman and A. Chaintreau (Eds.): NET-COOP 2008, LNCS 5425, pp. 70–77, 2009.

Our focus in this work is twofold. First, we focus on the conditional variance of LAS. Few results exist concerning the variance of LAS [11,2]. In [2], the asymptotic conditional variance is considered. In [11], the authors propose a classification of scheduling policies based on their variance. In particular, they propose to classify scheduling policies as: (i) always predictable, (ii) sometimes predictable or (iii) always unpredictable - precise definitions will be given in Section 2. LAS falls in the latter category, which seems at first sight disappointing. Indeed, LAS does a good job at providing low response time to small jobs but despite this nice property, the results in [11] seem to restrict the interest of LAS.

We revisit the variance of LAS by considering the case of an M/M/1/LAS queue. This case is interesting for two reasons. First, it is mathematically tractable. Second, the performance of LAS and especially the fraction of flows that receive a better service under LAS than under PS is known to increase with the variability of job size distribution [8]; and empirical distributions observed for Internet traffic have a much higher variability than the one of an exponential distribution [5].

Considering the M/M/1/LAS queue, we analytically bound the fraction of flows that are treated in a predictable manner. We obtain that at least 75% of flows are treated predicably, irrespectively of the load. Numerical studies further demonstrate that the actual fraction of such flows should be closer to 95%.

Second, we focus on the problem of using LAS queues in tandem. The motivation behind this scenario is to determine the benefits that could be obtained with LAS in a multiple bottlenecks scenario. A typical example is wireless mesh networks based on the 802.11 protocol where the available bandwidth is known to be highly varying, as has been exemplified by the roofnet experiment (http://pdos.csail.mit.edu/roofnet/doku.php). In such a situation, LAS could be highly beneficial as it allows to maintain a minimum level of interactivity, even when congestion is high. However, using LAS at multiple queues in tandem can also be detrimental to large flows that could be penalized multiple times.

To the best of our knowledge, no work has tackled the problem of studying LAS in tandem queues. We rely on numerical evaluations to address this issue as the analytical approach seems too complex at the moment, even for the case of M/M/1/LAS queues. We make the following contributions: (i) We demonstrate that while the departure process of an M/M/1/LAS queue is apparently a Poisson process, the LAS scheduling policy introduces a negative correlation between departures times and job sizes; (ii) We evaluate the impact of the above correlation on the conditional average response times of LAS queues in tandem.

2   Conditional Variance of an M/M/1/LAS Queue

The variance of the conditional response time for an $M/G/1/LAS$ queue with arrival rate λ is given by [12]:

$$Var[T(x)]^{LAS} = \frac{\lambda x \tilde{m}_2(x)}{(1 - \tilde{\rho}(x))^3} + \frac{\lambda \tilde{m}_3(x)}{3(1 - \tilde{\rho}(x))^3} + \frac{3}{4}\left(\frac{\lambda \tilde{m}_2(x)}{(1 - \tilde{\rho}(x))^2}\right)^2 \qquad (2.1)$$

Where $\{\tilde{m}_i(x)\}_{i \geq 1}$ are the truncated moments of the service time distribution. Truncated moments converge to the moments $\{m_i\}_{i \geq 1}$ when the job size tends to infinity. In particular, $\tilde{m}_1(\infty) = m_1 = \frac{1}{\mu}$ is the mean job service time. $\tilde{\rho}(x) = \lambda \tilde{m}_1(x)$ is the (truncated) load of the jobs up to size x, where $\tilde{\rho}(\infty) = \rho = \frac{\lambda}{\mu}$ is the load of the $M/G/1/LAS$ queue. Truncated moments for an exponential distribution $Exp(\mu)$ are given by:

$$\tilde{m}_i(x) = \int_0^x y^i \mu e^{-\mu y} dy + x^i e^{-\mu x}$$

In [11], scheduling policies were classified based on the variance of conditional response times as always predictable, always unpredictable or sometimes predictable. For a policy P, jobs of size x are treated predictably if:

$$\frac{Var[T(x)]^P}{x} \leq \frac{\lambda m_2}{(1 - \rho)^3} \tag{2.2}$$

Otherwise jobs of size x are treated unpredictably. See [11] for a justification of the right side term in Eq. (2.2). A policy is predictable for a given load ρ and a given service time distribution if all job sizes are treated predictably. More generally, a scheduling policy P is:

- **Always predictable.** If it is predictable under all loads and service distributions;
- **Sometimes predictable.** If it is predictable under some loads and service distributions, and unpredictable under others;
- **Always unpredictable.** If it is unpredictable regardless of service distribution and load.

In [11] a variance bound was derived to show that LAS is always unpredictable. In contrast, PS is shown to be always predictable. Comparison of LAS to PS is important as it is shown in [9] that the $M/G/1/LAS$ queue is an accurate model for a LAS router while the $M/G/1/PS$ queue is an accurate model for a FIFO router for connections with homogeneous RTT.

Our initial motivation in this paper is to show that LAS outperforms PS not only in terms of conditional response time offered to a majority of short jobs [8], but also in terms of conditional variance. Figure 1 illustrates the relative performance of LAS and PS for an $M/M/1$ queue with $\lambda = 1$ and $\mu = 1.25$, i.e. a load $\rho = 0.8$. We observe that while LAS offers both low average and variance of response time for most of the jobs, its performance becomes eventually worse than PS for the largest jobs.

The large variance observed for the large jobs in Figure 1 illustrates why LAS is classified as always unpredictable, but it is for the largest jobs only that Eq. (2.2) is violated. The question we address here is to determine the fraction of jobs that are treated predictably under LAS for the case of an $M/M/1/LAS$ queue.

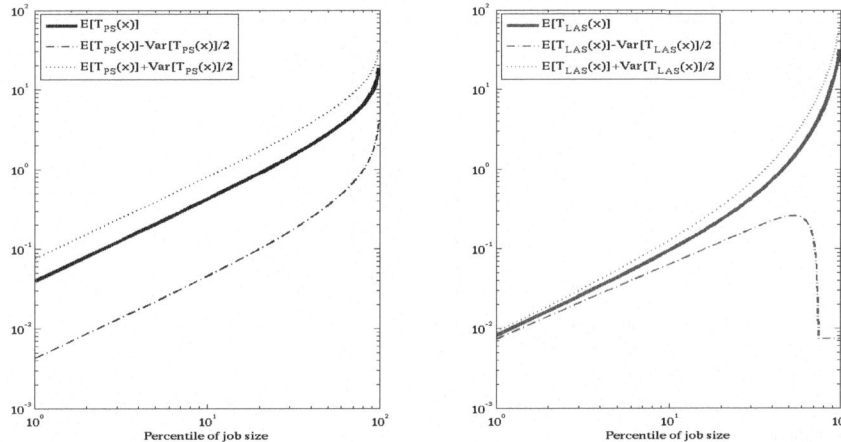

Fig. 1. M/M/1/LAS against M/M/1/PS: mean and variance of response time

2.1   M/M/1/LAS Variance Bounds

In this section, we first present two bounds on the variance of the conditional response time for an M/M/1/LAS queue. We next derive corresponding bounds on the fraction of jobs that are treated predictably under LAS. Due to space constraint, proofs are omitted, but can be found in our technical report [1].

Bound 1: For M/M/1/LAS,

$$Var[T(x)]^{LAS} \le \frac{\lambda x m_2}{(1 - \tilde{\rho}(x))^4}\left(1 + \frac{2}{e^2} - (\frac{1}{4} + \frac{2}{e^2})\tilde{\rho}(x)\right) \qquad (2.3)$$

Bound 2: For M/M/1/LAS,

$$Var[T(x)]^{LAS} \le \frac{8e\rho + 9\rho^2 - 8e\rho^2}{3e^2(1 - \rho)^4}x^2 \qquad (2.4)$$

Bounds 1 and 2 complement each other in the sense that bound 1 is more accurate for large job sizes while bound 2 is more accurate for small job sizes. Using these two bounds we further derive the percentage of jobs which are treated predictably in an M/M/1/LAS queue.

Bound 3: For M/M/1/LAS and a given ρ, at least a fraction of $\dfrac{1 - \left((1-\rho)^3(1 + \frac{2}{e^2})\right)^{1/4}}{\rho}$ of the jobs is treated predictably.

Bound 4: For M/M/1/LAS and a given ρ, at least a fraction of $1 - e^{-\frac{6e^2(1-\rho)}{8e + 9\rho - 8e\rho}}$ of the jobs is treated predictably.

Note that bounds 3 and 4 depend on ρ only, not on λ and μ. Bounds 3 and 4 show that the majority of jobs are in fact treated predictably in a $M/M/1/LAS$

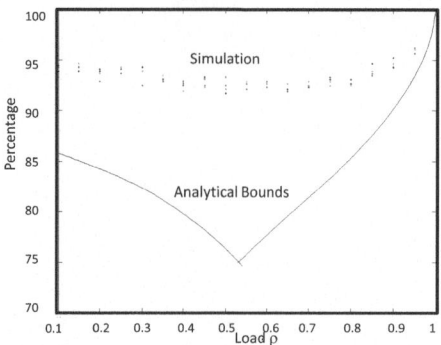

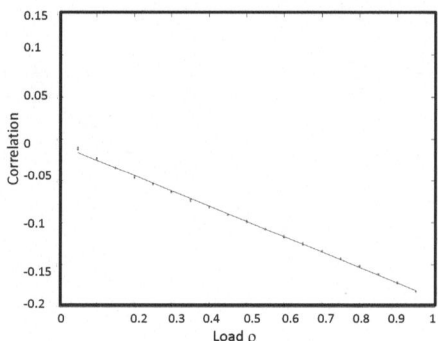

Fig. 2. Bounds 3 and 4 (solid line) and simulations (dots)

Fig. 3. Correlation of interdeparture times and jobsize for LAS

queue, as can be seen in Figure 2 where we plot the maximum of the two bounds (solid line). The minimum job percentage that is guaranteed to be treated predictably in this case, is about 75 percent.

As the above approach relies on lower bounding the number of jobs that are treated unpredictably under LAS, we further evaluated the percentage of jobs treated predictably using simulations. Figure 2 reports the results obtained by a simulator written in Matlab. Each simulation involved more than 500,000 jobs, and 4 different factorizations of ρ by using four different (λ, μ) pairs, which is why there are 4 points for each ρ value. We observe from Figure 2 that the fraction of flows that are treated predictably seems to depend on ρ only, not on the exact factorization of ρ. In addition, one can see that the actual percentage of jobs being treated predictably is very high: at least 90 percent of the jobs are treated predictably, regardless of the actual load value.

Note that for the case of Internet traffic, observed job size distributions are in general more skewed than the exponential one. As a consequence, we can expect that the fraction of jobs treated predictably is even higher than for the exponential case. As an illustration, for a Lognormal distribution with a coefficient of variation of 2, we observed that at all the load values we tested, between 10 and 90%, at least 99% of the jobs were treated predictably.

3   LAS in Tandem Queues

In this section, we focus on LAS in tandem queues where fresh arrivals occur only at the first queue and follow a Poisson process and service requirements follow an exponential distribution. We first study the output process of an isolated LAS queue and then, we will discuss the impact of the characteristics of the output process to the total response time of two LAS queues in tandem. Note that, to the best of our knowledge, no work has tackled this issue so far.

3.1   Characterization of the Output Process of an M/M/1/LAS Queue

LAS being a work-conserving and blind scheduling policy (job sizes are not known in advance), Burke's theorem is valid for an M/M/1/LAS queue. Hence, the output process for this queue is a Poisson process. We however show in the next section that job sizes and inter-departure times become correlated at the output of an M/M/1/LAS queue.

Correlation of Inter-departure Time and Job Size. To further characterize the output process of an M/M/1/LAS queue, we investigated whether job size is independent of interdeparture time. Figure 3 shows the correlation between job sizes and interdeparture times for LAS, plotted for different utilizations ρ. Again, different (λ, μ) pairs are used for the plots, but since curves overlap, correlation seems to depend on ρ only. Correlation is of negative type and increases with increasing load. We believe that this correlation stems from the fact that as load increases, large jobs are interrupted more frequently by shorter jobs. As a consequence, their remaining service requirement can reach very low values, lower than the service requirements of short jobs. When a large job leaves the queue, it is highly likely to be in a short period of time where load is low. During this period, many large jobs that were stuck in the queue, with small remaining service times, leave the system. We believe this explains the negative correlation we observe.

Note that we did not observe such a correlation for the PS queuing discipline (we do not present results here due to space constraint).

3.2   Impact on Tandem Queue Performance

The negative correlation observed above relates to the behavior of LAS that tends to sort jobs in the queue in ascending order, which means that short jobs tend to leave the LAS queue in groups and the same for large jobs. When considering the case of a tandem queues, this sorting introduced by the scheduler can have a detrimental impact on the performance of large flows. Indeed, when large flows reach the second queue, they again have to compete with the large flows they were in competition with in the first queue.

We quantitatively evaluated the previous intuition by comparing the conditional average response times[1] of two LAS queues in tandem with the sum of response time for the same system where we "re-draw" the job sizes of the job leaving the first queue using the initial distribution, thus wiping out the correlation observed in the previous section.

Figure 4 illustrates the above scenario for the case where load is 0.8. The left plot of Figure 4 represents the actual measured response time of the tandem system versus the added theoretical values (called independent queues in the graph). The right plot depicts the relative difference of the two systems.

---

[1] Results on the variance of the response time can be found in [1].

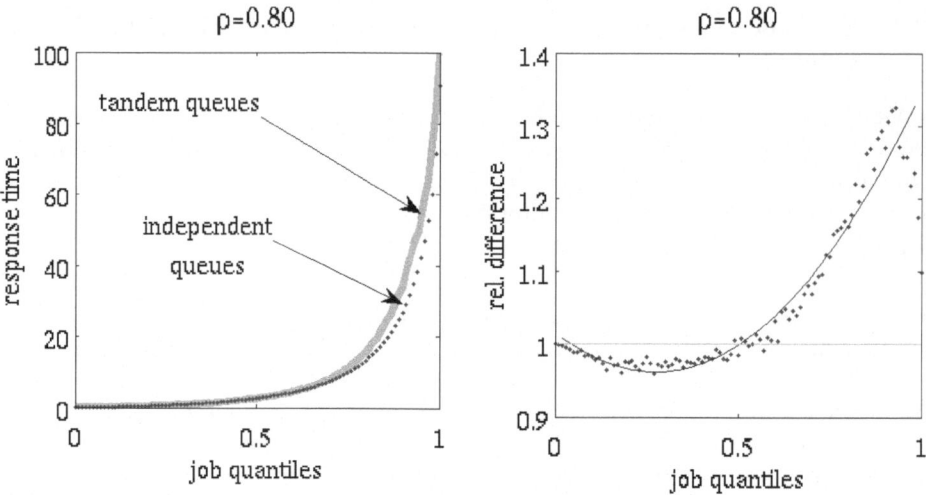

Fig. 4. Tandem Queue simulation results for LAS, $\rho = 0.80$

One can observe from Figure 4 that the observed negative correlation has a strong effect on the plots. While the response time is better than in the independent model for small jobs, LAS further penalizes large jobs in the tandem system.

4   Discussion

In this paper, we have focused both on the variance of the conditional response time of an M/M/1/LAS queue and the use of LAS in tandem queues.

For the variance, we proved that the fraction of jobs that are treated predictably is very high even in the unfavorable case of exponential service requirements. The practical implication of this result is that it is likely that when LAS is used for real Internet traffic, the variance in response time it offers to the majority of the flows be small as compared to the legacy FIFO scheduling policy.

Concerning the use of LAS in tandem queues, we demonstrated that while the output process of an M/M/1/LAS queue is Poisson, LAS tends to group jobs of similar sizes together, which results in a high penalty for the large jobs that cross the two queues. Practical implications of this result are less clear than for the variance case. Indeed routers work on a packet basis and do not output all the packets of a connection simultaneously as it happens in an M/M/1/LAS queue. This should dampen the effect we have observed.

As future work, we plan to continue working on the use of LAS in wireless mesh networks to precisely assess its performance. Note that a typical path in a wireless mesh network from an end user to an Internet gateway should be quite small, no more than 4 hops. As a consequence, we can expect that the nice properties of LAS, namely its ability to maintain interactivity even when the load is high, outweighs its side effect, namely the penalty experienced by large flows crossing multiple bottlenecks.

References

1. Auchmann, M., Urvoy-Keller, G.: On the variance of the least attained service policy and its use in multiple bottleneck networks. Technical Report RR-08-224, Institut Eurecom, France (June 2008),
   http://www.eurecom.fr/util/popuppubli.fr.htm?page=copyright&id=2497
2. Brown, P.: Comparing fb and ps scheduling policies. Technical report, Orange Labs (2008)
3. Crovella, M.E., et al.: A Practical Guide to Heavy Tails, ch. 3. Chapman and Hall, New-York (1998)
4. Harchol-Balter, M., Schroeder, B., Bansal, N., Agrawal, M.: Size-based scheduling to improve web performance. ACM Trans. Comput. Syst. 21(2), 207–233 (2003)
5. Hernandez-Campos, F., Karaliopoulos, M., Papadopouli, M., Shen, H.: Spatio-temporal modeling of traffic workload in a campus wlan. In: WICON 2006, p. 1 (2006)
6. Hu, M., Zhang, J., Sadowsky, J.: Size-aided opportunistic scheduling in wireless networks. In: GLOBECOM 2003. IEEE, Los Alamitos (2003)
7. Nuyens, M., Wierman, A.: The foreground-background queue: A survey. Perform. Eval. 65(3-4), 286–307 (2008)
8. Rai, I.A., Urvoy-Keller, G., Biersack, E.W.: Analysis of las scheduling for job size distributions with high variance. In: Proc. ACM SIGMETRICS, June 2003, pp. 218–228 (2003)
9. Rai, I.A., Urvoy-Keller, G., Vernon, M., Biersack, E.W.: Performance models for las-based scheduling disciplines in a packet switched network. In: ACM SIGMETRICS-Performance (June 2004)
10. Schrage, L.E.: The queue m/g/1 with feedback to lower priority queues. Management Science 13(7), 466–474 (1967)
11. Wierman, A., Harchol-Balter, M.: Classifying scheduling policies with respect to higher moments of conditional response time. ACM SIGMETRICS Performance Evaluation Review 33(1), 229–240 (2005)
12. Yashkov, S.: Processor-sharing queues: some progress in analysis. Queueing Systems: Theory and Applications 2(1), 1–17 (1987)

A Marginal Productivity Index Rule for Scheduling Multiclass Queues with Setups

José Niño-Mora*

Universidad Carlos III de Madrid
Department of Statistics
Avda. Universidad 30
28911 Leganés (Madrid), Spain
jnimora@alum.mit.edu
http://alum.mit.edu/www/jnimora

Abstract. This paper addresses the problem of designing a tractable scheduling rule for a multiclass $M/G/1$ queue incurring class-dependent linear holding costs and setup costs, as well as class-dependent generally distributed setup times, which performs well relative to the discounted or average cost objective. We introduce a new dynamic scheduling rule based on priority indices which emerges from deployment of a systematic methodology for obtaining marginal productivity index policies in the framework of restless bandit models, introduced by Whittle (1988) and developed by the author over the last decade. For each class, two indices are defined: an active and a passive index, depending on whether the class is or is not set up, which are functions of the class state (number in system). The index rule prescribes to engage at each time a class of highest index: it thus dynamically indicates both when to leave the class being currently served, and which class to serve next. The paper (i) formulates the problem as a semi-Markov multiarmed restless bandit problem; (ii) introduces the required extensions to previous indexation theory; and (iii) gives closed index formulae for the average criterion.

Keywords: Stochastic scheduling, optimal service control of queues, multiclass queues, setup times, setup costs, polling systems, index policies, marginal productivity index, queues with hysteresis.

1   Introduction

Consider a multiclass $M/G/1$ queue where the controller must dynamically allocate the server to one of M job classes, labeled by $m \in \mathbb{M} = \{1, \ldots, M\}$, which incurs linear holding costs at rate $h_m > 0$ per unit time per class m job in

* The author's research has been supported in part by the Spanish Ministry of Science and Innovation project MTM2007-63140 and an I3 faculty endowment grant, by the European Union's Network of Excellence Euro-FGI, and by the Autonomous Community of Madrid grant CCG07-UC3M/ESP-3389. The approach deployed herein was first announced by the author in a talk at the 13th INFORMS Applied Probability Conference (Ottawa, Canada, 2005).

E. Altman and A. Chaintreau (Eds.): NET-COOP 2008, LNCS 5425, pp. 78–86, 2009.

system. Switching the server from one class to another is costly: setting up the server so that it is ready to work on class m jobs entails an upfront setup cost $c_m \geq 0$, followed by a random setup time $D_m > 0$ having a general distribution with Laplace-Stieltjes transform (LST) $\gamma_m(\alpha) = \mathbb{E}\left[e^{-\alpha D_m}\right]$, with a finite second moment. At a given time, we say that the server is *on for class m* if it is either undergoing a setup or already set up for that class. Otherwise, we say that the server is *off for class m*. If the server is off for every class we say that it is *off* (for the system). The server can be turned on or off at *decision epochs* τ_k given by the instants of arrival when the server is off, along with the service and setup completion epochs. If the server is set up for a nonempty class, it is required to work in a job of that class. Yet, *idling* (cf. [1]) is allowed, meaning that the server can remain idle either while it is off for the system or while it is set up at an empty class, even though there may be jobs waiting at other classes. Class m jobs arrive as a Poisson process with rate λ_m, and their service times ξ_m have a general distribution with LST $\psi_m(\alpha) = \mathbb{E}\left[e^{-\alpha \xi_m}\right]$, having finite first and second moments $1/\mu_m$ and σ_m^2. Interarrival, service and setup times are mutually independent. The necessary stability condition $\rho < 1$ is assumed to hold, where $\rho \triangleq \sum_{m=1}^{M} \rho_m$ and $\rho_m \triangleq \lambda_m/\mu_m$.

Consider the problem of designing a scheduling policy $\boldsymbol{\pi}$, drawn from the space $\boldsymbol{\Pi}$ of nonanticipative randomized policies, which prescribes the action to take at each decision epoch, in order to minimize the expected total discounted cost relative to a given discount rate $\alpha > 0$, or the expected long-run average cost rate per unit time. A policy must resolve the following questions: (i) If the server is set up for a class, When should it leave (be turned off for) that class? (ii) When the server leaves a class, Should it be turned off for the system, or be turned on for another class? And, in the latter case, For which class? Denote by $a_m(\tau_k) \in \{0, 1\}$ the indicator of whether the server is on for class m at epoch τ_k, and let $a_m^-(\tau_k) \triangleq a_m(\tau_{k-1})$ denote the corresponding indicator for the previous epoch (where $a_m^-(\tau_0)$ indicates whether class m is initially set up, as $\tau_0 = 0$). Denote by $L_m(\tau_k)$ the number of class m jobs in system at epoch τ_k. Such processes are extended to piecewise constant right-continuous continuous time processes $a_m(t)$ and $L_m(t)$. The scheduling problems of concern are *semi-Markov decision processes* (SMDPs) having state and action processes $\mathbf{X}(\tau_k) = \left(X_m(\tau_k)\right)$ and $\mathbf{a}(\tau_k) = \left(a_m(\tau_k)\right)$, respectively, where $X_m(\tau_k) = \left(a_m^-(\tau_k), L_m(\tau_k)\right)$ is class m's *augmented state*. The discounted-cost problem is thus formulated as

$$\min_{\boldsymbol{\pi} \in \boldsymbol{\Pi}} \mathbb{E}_{\mathbf{x}}^{\boldsymbol{\pi}} \left[\sum_{m \in \mathbb{M}} \left\{ h_m \int_0^\infty L_m(t) e^{-\alpha t}\, dt + c_m \sum_{k=0}^\infty \left(1 - a_m^-(\tau_k)\right) a_m(\tau_k) e^{-\alpha \tau_k} \right\} \right],$$
(1)

while the average-cost problem is

$$\min_{\boldsymbol{\pi} \in \boldsymbol{\Pi}} \varlimsup_{T \to \infty} \frac{1}{T} \mathbb{E}_{\mathbf{x}}^{\boldsymbol{\pi}} \left[\sum_{m \in \mathbb{M}} \left\{ h_m \int_0^T L_m(t)\, dt + c_m \sum_{\tau_k \leq T} \left(1 - a_m^-(\tau_k)\right) a_m(\tau_k) \right\} \right], \quad (2)$$

where $\mathbb{E}_{\mathbf{x}}^{\boldsymbol{\pi}}[\cdot]$ denotes expectation under policy $\boldsymbol{\pi}$ conditioned on $\mathbf{X}(0) = \mathbf{x}$.

Researchers have extensively investigated the *performance analysis* of multi-class queues with setups under given policies, as such models are relevant in a variety of application areas, most notably in communication networks (*polling systems*) and in manufacturing systems. See, e.g., [2] and [3].

The *performance optimization* of multiclass queues with setups through dynamic scheduling has received substantially less attention, though interest has been steadily increasing over the last two decades. In [4], the special cases of (1) and (2) where there are two symmetric classes is investigated. It is shown that service at each queue should be exhaustive, and a threshold policy is proposed where the server switches from one empty class when the number of jobs in the other class exceeds a certain threshold. The design of optimal routing tables for polling in multiclass queues with setup times under the average criterion, relative to fixed policies for deciding when to leave a class, is addressed in [5,6]. In [7], an intuitive heuristic dynamic scheduling rule is proposed and assessed for the special case of (2) with only setup times. Such a rule is based on greedy indices defined for each class which represent reward rates, and which depend on whether the class is or not set up. The same authors use similar ideas to propose in [8] a rule for the corresponding problem with only setup costs. In [9], heavy-traffic Brownian approximations are used to design heuristic policies for the special cases of (2) where there are two classes and either only setup costs or times. Other heuristic policies are proposed in [10] and in [11]. The latter paper further proposed performance bounds based on fluid approximationss. Bounds based on mathematical programming relaxations are given in [12].

In contrast to the diversity of ad hoc heuristic ideas used in previous work, this paper introduces a novel approach to the design of dynamic scheduling policies for the above problems based on deployment of a unifying, systematic methodology. The latter applies to a versatile model termed the *multiarmed restless bandit problem* (MARBP), which concerns the optimal dynamic allocation of effort to multiple (semi-) Markovian *projects*, where at each time a project can be active or passive, resulting in different dynamics and/or cost rates. Problems (1)–(2) fit into such a framework by identifying the "projects" with job classes.

The MARBP was introduced in [13], as an extension of the classic *multiarmed bandit problem* (MABP). In the MABP, one project must be engaged at a time, and projects are assumed to be nonrestless, in that they do not change state while passive. In such a case, it was first proven by Gittins and Jones in [14] that there exists a priority *index* $\nu_m^*(i_m)$ attached to each project m as a function of its state i_m, such that the corresponding *index policy* that engages at each time a project of currently highest index is optimal. While the MARBP extension is generally intractable, [13] introduced a heuristic index rule based on indices $\nu_m^*(i_m)$ attached to restless projects. Yet, such an index only exists for a limited class of projects termed *indexable*. Over the last decade, the author has developed a body of theoretical and algorithmic work on restless bandit indexation, motivated by and illustrated on a variety of applications. See, e.g., [15,16,17,18,19] and the review paper [20]. Such work gives sufficient conditions for a restless project to be indexable, along with an adaptive-greedy algorithm for index computation, and furnishes an economically intuitive foundation for the indexation approach based on the concept of *marginal productivity index* (MPI), which unifies and extends previous indices proposed in the literature. The incorporation of switching penalties in bandit models has been addressed by the author in classic bandits [19,21] and in restless bandits [22].

Yet application of such an approach to the model of concern is not straightforward, as it requires significant extensions of previous results —which applied either to finite-state projects or to countable-state projects with a linearly ordered state space— whereas the projects herein have a countable partially ordered state space. This paper outlines the required extensions, and shows how to deploy them to obtain new dynamic index policies for problems (1) and (2).

The remainder of the paper, which focuses for concreteness in the discounted problem (1), is organized as follows. Section 2 formulates the problem as a MARBP, and outlines the indexation approach to the design of a dynamic index policy. Section 3 introduces the required extensions of previous results. Section 4 deploys such results to obtain the new index rule, which is given in closed form for the average-cost problem. Finally, Section 5 ends the paper with some concluding remarks.

Due to space constraints no proofs are given. These will be included in the full paper version, along with the results of an extensive computational study.

2   Restless Bandit Indexation Approach

In the MARBP reformulation of the problem of concern, project m has state $X_m(t) = \big(a_m^-(t), L_m(t)\big)$. Consider now the subsystem consisting of class m *in isolation*, as a controlled queue in which the server can be turned on or off at each decision epoch. Let us introduce a parameter ν which represents the *wage rate* to be paid to the server per unit time that it is on, and consider the optimal dynamic control subproblem for such a queue:

$$
\min_{\pi_m \in \Pi_m} \mathbb{E}^{\pi_m}_{(a_m^-, i_m)} \left[\left\{ \int_0^\infty \{h_m L_m(t) + \nu a_m(t)\} e^{-\alpha t}\, dt \right. \right.
$$
$$
\left. \left. + c_m \sum_{k=0}^\infty \big(1 - a_m^-(\tau_k)\big) a_m(\tau_k) e^{-\alpha \tau_k} \right\} \right],
\tag{3}
$$

where Π_m is the space of nonanticipative server-operating policies, and $\mathbb{E}^{\pi_m}_{(a_m^-, i_m)}[\cdot]$ denotes expectation under policy π_m conditioned on $X_m(0) = (a_m^-, i_m)$.

Note that several variations of problem (3), typically focusing on its time-average counterpart and assuming only setup costs, have been extensively investigated in the literature on optimal control of queues. See, e.g., [23,24,25,26,27,28].

We allow the server's wage ν to take any real value, and investigate the structure of optimal policies for (3) obtained by varying it as a parameter $\nu \in \mathbb{R}$. Let us say that problem (3) is *indexable* if the set of states (a_m^-, i_m) where it is optimal to take the active action $a_m = 1$ at a decision epoch decreases monotonically from the full state space $\mathbb{X}_m \triangleq \{0,1\} \times \mathbb{Z}_+$ to the empty set $\emptyset$ as the server's wage ν increases from $-\infty$ to ∞, where $\mathbb{Z}_+ \triangleq \{0,1,2,\ldots\}$. In such a case, optimal policies for (3) are characterized by an *index* $\nu_m^*(a_m^-, i_m)$ for $(a_m^-, i_m) \in \mathbb{X}_m$, which we will term the class' *marginal productivity index* (MPI): it is optimal to take action $a_m = 1$ in state (a_m^-, i_m) iff $\nu_m^*(a_m^-, i_m) \geq \nu$. Such an index measures the marginal productivity of the server's effort in each state, as elucidated in the economic interpretation in [17]. Note that the index decouples into an *active index* $\nu_m^*(1, i_m)$, which applies when the server is set up, and a *passive index* $\nu_m^*(0, i_m)$, which applies otherwise.

Assume now that each class m is indexable, having MPI $\nu_m^*(a_m^-, i_m)$. The economic interpretation of such indices suggests the following (priority-) index rule for the multiclass scheduling problem (1). Suppose that at a decision epoch τ_k class m occupies state $X_m(\tau_k) = (a_m^-, i_m)$, for $m \in \mathbb{M}$, and consider the index values $\nu_m^*(a_m^-, i_m)$ at such states. If none of them is positive, i.e., if $\nu_m^*(a_m^-, i_m) < 0$ for $m \in \mathbb{M}$, then turn or keep off the server. Otherwise, pick a class m^* which attains the largest index value among those that are positive at the current state, and turn or keep the server on for such a class.

Intuition suggests that such indices should satisfy the *hysteretic property*

$$\nu_m^*(1, i_m) \geq \nu_m^*(0, i_m), \quad i_m \geq 0, \tag{4}$$

so, other things being equal, one should keep the server in a class where it is set up rather than switching and setting it up for another class.

3   Outline of Required Extension of Indexation Theory

In order to deploy the indexation approach outlined above we need to (i) establish that the projects of concern are indexable, and (ii) give a tractable approach to compute the index. This turns out to require significant extensions of previous results on restless bandit indexation introduced by the author. We next outline such extensions as they apply to the model of concern.

Consider an individual project representing a single queue with setups subject to service control, where we drop henceforth the class label m from the notation. In such a setting, we evaluate the value of a policy π for operating such a single-server queue using the discounted *cost measure*

$$f_{(a^-,i)}^\pi \triangleq \mathbb{E}_{(a^-,i)}^\pi \left[\int_0^\infty h L(t) e^{-\alpha t}\, dt + c \sum_{k=0}^\infty \left(1 - a^-(\tau_k)\right) a(\tau_k) e^{-\alpha \tau_k} \right],$$

and further evaluate the server's work expended (where thet time spent in setups is counted as work) using the discounted *work measure* (cf. [16])

$$g_{(a^-,i)}^\pi \triangleq \mathbb{E}_{(a^-,i)}^\pi \left[\int_0^\infty a(t) e^{-\alpha t}\, dt \right].$$

We can thus formulate problem (3) as

$$\min_{\pi \in \Pi} f_{(a^-,i)}^\pi + \nu g_{(a^-,i)}^\pi. \tag{5}$$

We will refer to (5) as the *ν-wage problem*, as it concerns finding an optimal server-operating policy that minimizes the sum of holding, switching and server's labor costs. We can then use such a parametric problem to define the concept of *indexability* and MPI $\nu^*(a^-, i)$ as in the previous section.

Now, the key to achieving goals (i) and (ii) above is to exploit special structure, by being able to *guess* a family of stationary deterministic policies among which an optimal policy for (5) exists for any wage $\nu \in \mathbb{R}$. We will any such policy by its *active (state) set*, which is the set of states where the policy prescribes to take the active action at a decision epoch. Such an active set is conveniently represented by giving two sets $S_0, S_1 \subseteq \mathbb{Z}_+$, with the following interpretation: in state $(0, i)$, turn the server on iff $i \in S_0$; in state $(1, i)$, keep the server on iff $i \in S_1$. We will also denote by $S_0 \oplus S_1$ the corresponding policy.

Intuition suggests that it suffices to consider *hysteretic* policies given by a passive (k) and and active (l) threshold satisfying $k \geq l \geq 0$, so $S_0 = \{k, k+1, \ldots\}$ and $S_1 = \{l, l+1, \ldots\}$, as well as policies where $S_0 = \emptyset$ and $S_1 = \{l, l+1, \ldots\}$, where we take $k = \infty$. We will also denote by (k, l) the corresponding policy, writing, e.g., $g_{(a^-, k)}^{(k,l)}$. We will denote by $\mathcal{F}$ the family of all such active sets, and will refer to the family of $\mathcal{F}$-*policies*.

Let us say that problem (5) is $\mathcal{F}$-*indexable* if (i) it is indexable; and (ii) for every wage $\nu \in \mathbb{R}$ there is an optimal $\mathcal{F}$-policy for (5). Note that, in such a case, the MPI $\nu^*_{(a^-, i)}$ satisfies the hysteretic property (4).

In the finite-state case, previous indexation theory (cf. [20]) prescribes to use the *adaptive-greedy index algorithm* introduced in [15] to compute the MPI. In the present setting, however, the fact that the augmented state space $\mathbb{X} \triangleq \{0, 1\} \times \mathbb{Z}_+$ is countable yet not linearly ordered calls for a suitable extension, which we introduce next as follows. We must identify a *linear ordering* $\preceq$ of the state space $\mathbb{X}$ which satisfies certain conditions, where we write

$$\mathbf{S}_x \triangleq \{y \in \mathbb{X} \colon x \prec y\}, \quad x \in \mathbb{X}.$$

Such conditions are: (i) $\mathbf{S}_x \in \mathcal{F}$ for every $x \in \mathbb{X}$; (ii) the work measure is monotone consistently with such an ordering, in that $\left(g_y^{\mathbf{S}_{x'}}\right)_{y \in \mathbb{X}} \succsim \left(g_y^{\mathbf{S}_x}\right)_{y \in \mathbb{X}}$ if $x \prec x'$; and (iii) the *marginal productivity rates*

$$\nu^*(x) \triangleq \frac{f_x^{\mathbf{S}_x} - f_x^{\mathbf{S}_x \cup \{x\}}}{g_x^{\mathbf{S}_x \cup \{x\}} - g_x^{\mathbf{S}_x}}$$

are monotone consistently with the ordering $\preceq$, i.e., $\nu^*_x \leq \nu^*_{x'}$ if $x \preceq x'$. Instead of (iii), one can use the alternative condition (iii'): $\mathcal{F}$-policies are optimal for problem (5) under any wage $\nu \in \mathbb{R}$.

Theorem 1. *Under the stated conditions, the restless project is $\mathcal{F}$-indexable and $\nu^*(x)$ is its MPI.*

4   Application to the Scheduling Model

To deploy the above results in the model of concern it must be established that the required conditions (i)—(iii) in Theorem 1 hold relative to a certain linear ordering of $\mathbb{X}$. The following result ensures that such is indeed the case.

Lemma 1. *There exists a linear ordering $\preceq$ on $\mathbb{X}$ satisfying conditions (i)—(iii) in Theorem 1, under which $(0,i) \prec (1,i)$ and $(a^-,i) < (a^-,i+1)$ for $(a^-,i) \in \mathbb{X}$.*

The required ordering $\preceq$ can be readily constructed through a suitable modification of the adaptive-greedy algorithm referred to above, which is not included here due to space constraints.

Further, by taking appropriate Maclaurin expansions as the discount factor α vanishes, one obtains a corresponding index rule for the (long-run) average criterion. Under the latter, we can actually obtain closed-form formulae for the index. In such a case, the nested decreasing active-set ordering corresponding to the state ordering $\preceq$ has the form

$$(1,0),\dots,(k^*-1,0),(k^*,0),(k^*,1),(k^*+1,1),(k^*+2,1),\dots,(\infty,1),(\infty,2),\dots$$

where k^* is a critical threshold value given by

$$k^* = \max_{k \geq 1} \left\{ c\left(\frac{\lambda}{k} + \frac{1}{\mathbb{E}[D]}\right) - h\frac{(k+\lambda\mathbb{E}[D])\left((k-1)k - \lambda^2\mathbb{E}[D^2]\right)}{2k\lambda(1-\rho)\mathbb{E}[D]} > 0 \right\}.$$

We obtain that the discounted MPI of state $(1,i)$, for $i \geq 1$, has the Maclaurin expansion

$$\nu^*(1,i) = \frac{h\mu}{\alpha} + O(1), \quad \text{as } \alpha \searrow 0.$$

Further, the MPI of state $(1,0)$ is given by

$$\nu^*(1,0) = c\frac{\lambda}{k^*} + h\frac{(k^*-1)(k^*+2\lambda\mathbb{E}[D]) + \lambda\left(2\mathbb{E}[D] + \lambda\mathbb{E}[D^2]\right)}{2k^*(1-\rho)} + O(\alpha).$$

The MPI of state $(0,i)$, for $i \geq k^*$ is

$$\nu^*(0,i) = h\frac{i(1+2\lambda\mathbb{E}[D]) + i^2 + \lambda^2(\mathbb{E}[D]^2 - \text{Var}[D])}{2\lambda(1-\rho)\mathbb{E}[D]} - \frac{c}{\mathbb{E}[D]} + O(\alpha),$$

and, for $0 \leq i < k^*$,

$$\nu^*(0,i) = h\frac{i + \lambda\mathbb{E}[D]}{1-\rho} + O(\alpha).$$

Note that in the multiclass model, for a discount rate α small enough, and two distinct classes m, n, with $i_n \geq 1$, it holds that $\nu_n^*(1,i_n) > \nu_m^*(0,j_m)$, the MPI scheduling rule is *exhaustive*. Other qualitative properties of such a rule are readily obtained from the above formulae, which are consistent with the properties of optimal policies established under symmetry assumptions in [29].

Such index formulae are valid for the case of positive setup times $D > 0$. In the case where there are no setup times yet there are positive setup costs, a similar analysis can be carried out which yields different index formulae.

5   Conclusions

This paper introduces a novel approach to the design of tractable dynamic scheduling policies of priority-index type for multiclass queues with both setup costs and times, based on formulating such models as multiarmed restless bandit problems and then deploying the general theoretical and algorithmic results available for the latter, suitably extended to the present setting. A full version of this paper will report on an experimental study to assess both the degree of suboptimality of the resultant index rules and their relative performance compared to other policies previously proposed in the literature.

References

1. Eisenberg, M.: The polling system with a stopping server. Queueing Syst. 18, 387–431 (1994)
2. Takagi, H.: Analysis of Polling Systems. MIT Press, Cambridge (1986)
3. Levy, H., Sidi, M.: Polling systems: Applications, modelling and optimization. IEEE Trans. Comm. 38, 1750–1760 (1990)
4. Hofri, M., Ross, K.W.: On the optimal control of two queues with server setup times and its analysis. SIAM J. Comput. 16, 399–420 (1987)
5. Boxma, O.J., Levy, H., Weststrate, J.A.: Efficient visit frequencies for polling tables: Minimization of waiting cost. Queueing Syst. 9, 133–162 (1991)
6. Boxma, O.J., Levy, H., Weststrate, J.A.: Efficient visit frequencies for polling systems. Performance Evaluation 18, 103–123 (1993)
7. Duenyas, I., Van Oyen, M.: Heuristic scheduling of parallel heterogeneous queues with setups. Management Sci. 42, 814–829 (1996)
8. Duenyas, I., Van Oyen, M.: Stochastic scheduling of parallel queues with setup costs. Queueing Syst. 19, 421–444 (1995)
9. Reiman, M.I., Wein, L.M.: Dynamic scheduling of a two-class queue with setups. Oper. Res. 46, 532–547 (1998)
10. Olsen, T.L.: A practical scheduling method for multiclass production systems with setups. Management Sci. 45, 116–130 (1999)
11. Lan, W.M., Olsen, T.L.: Multiproduct systems with both setup times and costs: fluid bounds and schedules. Oper. Res. 54, 505–522 (2006)
12. Bertsimas, D., Niño-Mora, J.: Optimization of multiclass queueing networks with changeover times via the achievable region approach: Part I, the single-station case. Math. Oper. Res. 24, 306–330 (1999)
13. Whittle, P.: Restless bandits: Activity allocation in a changing world. In: Gani, J. (ed.) A Celebration of Applied Probability; J. Appl. Probab. Applied Probability Trust 25A (spec. vol.), 287–298, Sheffield, UK (1988)
14. Gittins, J.C., Jones, D.M.: A dynamic allocation index for the sequential design of experiments. In: Gani, J., Sarkadi, K., Vincze, I. (eds.) Progress in Statistics (European Meeting of Statisticians, Budapest, 1972), pp. 241–266. North-Holland, Amsterdam (1974)
15. Niño-Mora, J.: Restless bandits, partial conservation laws and indexability. Adv. Appl. Probab. 33, 76–98 (2001)
16. Niño-Mora, J.: Dynamic allocation indices for restless projects and queueing admission control: a polyhedral approach. Math. Program. 93, 361–413 (2002)

17. Niño-Mora, J.: Restless bandit marginal productivity indices, diminishing returns and optimal control of make-to-order/make-to-stock $M/G/1$ queues. Math. Oper. Res. 31, 50–84 (2006)
18. Niño-Mora, J.: Marginal productivity index policies for scheduling a multiclass delay-/loss-sensitive queue. Queueing Syst. 54, 281–312 (2006)
19. Niño-Mora, J.: A faster index algorithm and a computational study for bandits with switching costs. INFORMS J. Comput. 20, 255–269 (2008)
20. Niño-Mora, J.: Dynamic priority allocation via restless bandit marginal productivity indices. TOP 15, 161–198 (2007)
21. Niño-Mora, J.: Computing an index policy for bandits with switching penalties. In: ValueTools 2007: Proceedings of the Second International Conference on Performance Evaluation Methodologies and Tools (Nantes, France). ACM International Conference Proceedinsg Series, ICST, Brussels, Belgium (2007)
22. Niño-Mora, J.: Marginal productivity index policies for scheduling restless bandits with switching penalties. In: Albers, S., Möhring, R.H., Pflug, G.C., Schultz, R. (eds.) Algorithms for Optimization with Incomplete Information. Number 05031 in Dagstuhl Seminar Proceedings (2005)
23. Yadin, M., Naor, P.: Queueing systems with a removable service station. Oper. Res. 14, 393–405 (1963)
24. Bell, C.E.: Characterization and computation of optimal policies for operating an $M/G/1$ queuing system with removable server. Oper. Res. 19, 208–218 (1971)
25. Bell, C.E.: Optimal operation of an $M/G/1$ priority queue with removable server. Oper. Res. 21, 1281–1290 (1973)
26. Heyman, D.: Optimal operating policies for $M/G/1$ queuing systems. Oper. Res. 16, 362–382 (1968)
27. Sobel, M.J.: Optimal average-cost policy for a queue with start-up and shut-down costs. Oper. Res. 17, 145–162 (1969)
28. Borthakur, A., Medhi, J., Gohain, R.: Poisson input queueing system with startup time and under control-operating policy. Comput. Oper. Res. 14, 33–40 (1987)
29. Liu, Z., Nain, P., Towsley, D.: On optimal polling policies. Queueing Syst. 11, 59–83 (1992)

A Heuristic Approach to the Passive Detection of Reno-Like TCP Flows*

Miguel Rodríguez-Pérez, Manuel Fernández-Veiga, Sergio Herrería-Alonso, and Cándido López-García

E.T.S.E. Telecomunicación
Campus universitario s/n
36 310 Vigo, Spain

Abstract. Traditional TCP-Reno like congestion control protocols exhibit poor performance when deployed in fast or very large network paths. Delay based congestion avoidance mechanisms (DCA), like FAST-TCP, get much higher performance, in the same circumstances. However, when mixed with TCP-Reno or alike traffic they are unable to attain their fair share of bandwidth. In this paper we present a new mechanism that can indirectly detect the present of non DCA-friendly traffic that can be used by new DCA algorithms to auto-tune themselves with more aggressive parameters to achieve their fair share.

Keywords: Congestion control, Delay-based, FAST-TCP, TCP-Vegas.

1 Introduction

TCP congestion control protocols are having problems in keeping fully utilized current high-speed links. Very high-throughput implies very big congestion window sizes, which in turns needs negligible packet losses. For protocols that just employ packet losses as a signal of congestion, increasing the bandwidth means a substantial waste of resources if transmission errors were confused with packet losses really caused by congestion. Hence, new congestion control mechanisms are needed to fully use the installed network capacity under this scenario. The research community has come up with new protocols more aggressive than Reno to compensate for the lack of feedback that packet losses provide [1,2]. Another promising approach consist in adapting the techniques of Delay-based congestion avoidance (DCA) [3,4,5] protocols that gather more information feedback from the net, and thus should be able to better allocate its resources.

Delay-based congestion avoidance (DCA) control protocols, and FAST-TCP [5] in particular, are known for their effectiveness when used in isolation: low queueing delay, convergence to a constant transmission rate, low packet losses, full utilization of the links bandwidth, ... [6,4] However, when confronted against

---

* This work was supported by the "Ministerio de Educación y Ciencia" through the project TSI2006-12507-C03-02 of the "Plan Nacional de I+D+I" (partly financed with FEDER funds).

E. Altman and A. Chaintreau (Eds.): NET-COOP 2008, LNCS 5425, pp. 87–94, 2009.

Reno traffic they overreact to congestion, leading to an unnecessarily poor performance [6]. This has prevented wide-scale deployment of these kind of protocols despite being available, although not enabled by default, in the Linux kernel since at least 2004 for Vegas.

Thus, in the absence of fair-queueing mechanisms in the network, the ability to reliably and rapidly detect the presence of loss-reacting congestion control protocols in a network path is key for the successful deployment of DCA protocols in the Internet.

Because in a network with only DCA flows packet losses should be really scarce, the existence of quasi-periodic packet losses can be taken as an indication of the existence of Reno-like TCP flows. However, DCA sources cannot usually detect directly the losses, because dropped packets are more likely to belong to Reno flows. This happens for two reasons: firstly, Reno flows have a much higher chance of suffering losses because they get a higher share of the bandwidth, and send more packets per second. The second is that, as the queue gets full, DCA flows decrease their sending rate due to the increased queueing delay, so it is just when a link is closest to congestion that DCA flows throttle their traffic, decreasing even more the chance that the dropped packet will be theirs. In the evaluation section of this paper we show some scenarios in with FAST flows get no direct losses, while Reno flows share a common bottleneck.

In this paper we propose a simple but effective mechanism to detect the presence of loss-based congestion control algorithms (like Reno and its variants) in a given network path. This proposal permits the creation of new dual algorithms that are able to rapidly adapt to the presence of Reno flows. Our proposal tries to look at the effects of AIMD traffic in the network and, as such, predicts episodes of packet losses in the network caused by the tendency of Reno variants to fill the router queues. Compared with [7], we can detect and thus react to Reno flows in the timescale of the RTT and, in contrast to [8], we do not mistake the presence of many DCA flows with Reno flows, unless the network is under-provisioned. But in that case DCA algorithms behave almost exactly like Reno.

The rest of this paper is organized as follows. Section 2 provides information about DCA algorithms and their compatibility with Reno. Section 3 presents the basis of our passive loss detection algorithm. Then, in Section 4 representative results from experimental evaluation are shown. The conclusions are in Section 5.

2 Background

DCA algorithms work on the assumption that it is possible to infer the network status by observing the variations in the RTT. The difference between the RTT and the propagation delay is directly related to the amount of data in transit, this amount being a measure of the level of congestion. So, adjusting the window size based on these variations, DCA flows keep an appropriate transmission rate without causing congestion. In contrast, TCP-Reno and its variants need to drive the network close to congestion in order to receive the feedback needed to adjust the window. The window size is slowly increased until it reaches a point that buffers overflow and a packet loss occurs. This leads sources to abruptly reduce the sending rate and the slow increment begins again. This prevents Reno flows to fully use all the available bandwidth. Thus, DCA algorithms are more suitable in long fat pipes where packet losses are too scarce to properly adjust the rate or for those applications negatively affected by sudden changes in the transmission rate.

Both TCP-Vegas and FAST-TCP employ a similar *modus operandi*, in fact, FAST-TCP can be treated as an improved (faster) version of the former [5].

To adjust its window size w in the absence of packet losses FAST-TCP sources keep an estimation of both the current RTT, r, and the propagation delay r_0, approximated in practice as the minimum RTT measured throughout the connection lifetime. Then, w is periodically updated according to

$$w \longleftarrow \gamma \left(\frac{r_0 w}{r} + \alpha \right) + (1 - \gamma)w. \tag{1}$$

Here, α and γ are configuration parameters: α plays the same role that α and β play in Vegas, representing the amount of data enqueued in the network when stability is reached, while γ controls the speed of convergence and is usually set to 0.5. Several papers deal with the conditions that both α and γ must met to reach equilibrium. More details can be found in [9,10,11,12].

In comparison, TCP Reno and its variants induce packet losses to estimate the available bandwidth in the network. To do so, they slowly increase their transmission rates up to the point of filling network buffers. When this happens, the transmission rate is halved and the cycle repeats again.

Although Reno behavior leads to fair bandwidth share between Reno flow and network stability, it causes an unfair bandwidth share against DCA protocols. In fact, according to [13] if $k < \alpha$ is the number packets DCA flows have enqueued at a given bottleneck of capacity B packets, then the relative throughput of DCA flows is given by

$$\rho = \frac{2k}{B - k}. \tag{2}$$

If the buffer is overprovisioned, which is usually the case, $B \gg k$ and DCA flows are unable to attain their fair share.

In the last couple of years there have started to appear new hybrid congestion control proposals that try to combine the behavior of DCA flows with the aggressiveness of Reno variants depending on the network state [7,8]. In [7] the authors propose a way to adapt FAST-TCP α parameter that takes into account both loss rate and queueing delay. However, the configuration is updated only once every two minutes, leading to slow reaction and long convergence times. Alternatively, the authors of [8] use queueing delay to decide whether to switch between DCA or AIMD algorithms, employing DCA when the measured queueing-delay is sufficiently small. The drawback is that they also unnecessarily employ AIMD algorithms when there are many DCA flows in the network path, degrading overall performance.

3 Procedure

In a properly dimensioned network shared only by delay-based flows the equilibrium point is maintained with a complete lack of packet losses caused by congestion. On the contrary, loss-reacting congestion control protocols —like TCP-Reno or its variants— cannot reach a complete equilibrium and cause periodic packet losses. With this in mind, a FAST-TCP flow could look for packet losses to infer the presence of non DCA-friendly TCP flows.

However, the unfair sharing of a bottleneck between flows employing different congestion control algorithms makes FAST-TCP less likely to suffer from direct packet losses that the Reno-like flows, so a method to detect packet losses, even if not own, is needed. We provide simulation scenarios later in this paper that show this phenomenon.

In a bottleneck shared by TCP-Reno flows, the queue occupation follows cyclic patterns, first getting filled up while flows struggle to reach the highest sustainable throughput, to be drained later while traversing flows halve their windows to adapt to congestion. Thus, an observer could detect packet losses monitoring the dynamics of the queue length.

Queue length is precisely what FAST-TCP flows indirectly monitor to react to congestion. In fact, they estimate their own queueing delay to maintain α packets enqueued along its network path by measuring both the round-trip-time (r) and the round-trip propagation delay (r_0). These data can be reused to estimate not only its own queueing delay, but the total delay caused by all flows sharing the network path. In fact, the difference $r - r_0$ is directly proportional to the amount of data enqueued at the bottleneck link.

Let $r(n)$ be the round trip time measured by a FAST-TCP flow after the arrival of acknowledgment n. Then, the queueing delay can be calculated as $d(n) = r(n) - r_0$. To avoid having all values dependant on the network characteristics and packet length, define $\bar{q}(n) \triangleq \frac{d(n)}{r_0}$ as the normalized queueing delay (which is clearly proportional to the normalized queue length). Lastly, let $\Delta\bar{q}(n) = \bar{q}(n) - \bar{q}(n-1)$ be the per-packet variation of the queue length.

The ideal loss signal

$$l(n) = \begin{cases} 1 & \text{if losses occurred between packets } n-1 \text{ and } n \\ 0 & \text{in any other case,} \end{cases} \tag{3}$$

can be tried to be recovered using an heuristic based on the variations of the queueing delay as

$$\tilde{l}(n) = \begin{cases} 1 & \text{if } \frac{-\Delta\bar{q}(n)}{\bar{q}(n-1)} > \tau \\ 0 & \text{in any other case.} \end{cases} \tag{4}$$

Equation (4) merits several comments. We deduce that there must have been packet losses when the decrement of queueing delay represents a significant portion of the total queueing delay. It is important to note that the value of the threshold τ should be chosen carefully. There is certainly a tradeoff between a too low value that could yield many false positives, for instance when a flow leaves the network, and a too high value that would miss many losses.

4   Experimental Results

Throughout this section we will present some of the results we have obtained testing the accuracy of our estimated loss signal. Recall that it is the ability to detect the presence of AIMD flows what we are interested in testing, and not the detection of every packet loss event. To this end, we have modified the FAST-TCP implementation for the ns-2 [14] simulator so that it is able to report the recovered $\hat{l}(n)$ vector.

For our first set of experiments we used a topology like that shown in Fig. 1. There is a set of n_F FAST-TCP flows competing for bandwidth with n_R Reno flows. Unless otherwise stated, all FAST-TCP flows have been set up with $\alpha = 3$ and network buffers with capacity for $B = (\alpha+1)n_F+10n_R$ packets. The decision threshold has been set experimentally to a 20 percent reduction ($\tau = 0.2$). All the flows carry a endless stream of application data from S_i towards D_i.

Our first experiment measures the time needed to detect a given packet loss. We show the results for $n_F = 1$ and $n_R = [1, 50]$ in Fig. 2. The propagation delay in the links $S_i \rightarrow R_1$ has been modified so as to make round-trip propagation delay of the FAST-TCP flow one tenth, equal and ten times as large as the one of the competing Reno flows. As the figure shows, it takes about two RTT for loss detection, which is quite fast indeed, given that what is used for detection is the reaction of Reno flows to packet losses.

Our second experiment shows both $l_i(n)$ and $\hat{l}_F(n)$. Fig. 3 represents with a little mark packet losses suffered by Reno flows, and by dashed lines the losses detected by a modified FAST-TCP flow. Each TCP flow is assigned an unique sequential identifier to help to identify them in the y-axis. Again, we modify link propagation delays so as to compare the results for different ratios of the RTT of the Reno flows against that of FAST-TCP. The results are for fifty Reno flows

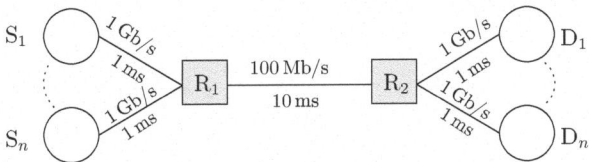

Fig. 1. Network test bed

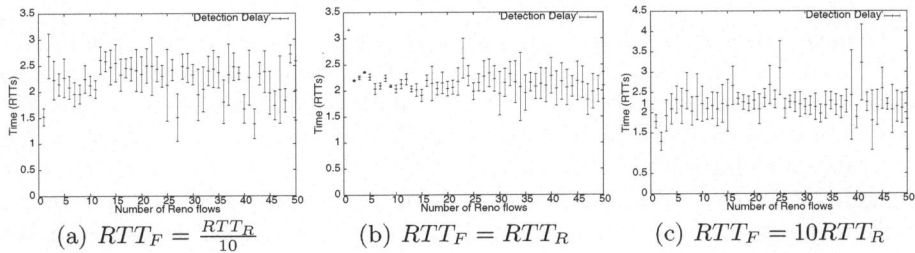

(a) $RTT_F = \frac{RTT_R}{10}$ (b) $RTT_F = RTT_R$ (c) $RTT_F = 10RTT_R$

Fig. 2. Average delay in detection of a packet loss for different propagation delays

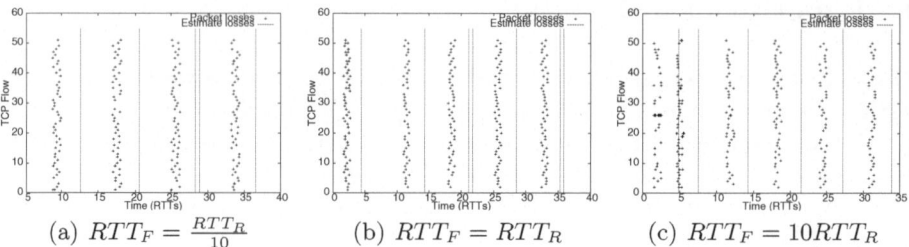

(a) $RTT_F = \frac{RTT_R}{10}$ (b) $RTT_F = RTT_R$ (c) $RTT_F = 10RTT_R$

Fig. 3. Loss pattern for a dumb-bell topology with 50 Reno flows for different propagation delays

and a single FAST-TCP flow. Observe how losses are heavily correlated between TCP flows, and how, despite the appearance of some false positives (RTT 28 in Fig. 3(a) and RTTs 22 and 36 in Fig. 3(b), for instance), those are always a double positive and not a false indication of packet losses.

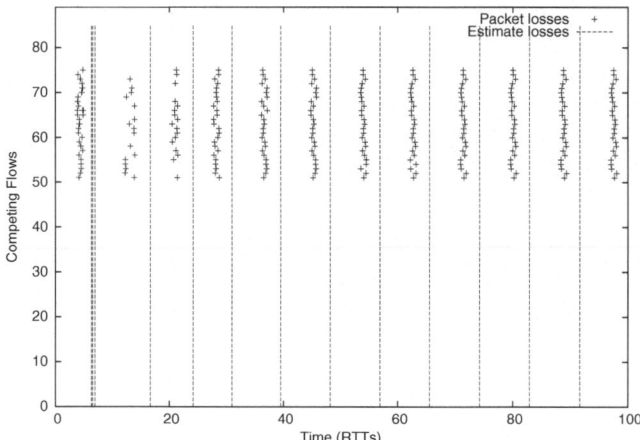

Fig. 4. Loss pattern for a dumb-bell topology with 50 FAST-TCP flows for 25 number of Reno flows

We have repeated the above experiments, but this time employing fifty FAST-TCP flows competing against one, 25 and 50 Reno flows. The results for the 25 Reno flows case appear in Fig. 4. FAST-TCP flows have ids between one and fifty while Reno flows have ids from 51 onwards. We have only run the loss detection algorithm in a single FAST-TCP flow.

Again, the recovered signal $\hat{l}(n)$ follows closely the measured $l(n)$ for any of the Reno flows. It is also important to note that the FAST-TCP flows get almost no losses, so, in absence of a loss detection algorithm they would be unable to discern whether they are competing with more aggressive AIMD flows and react

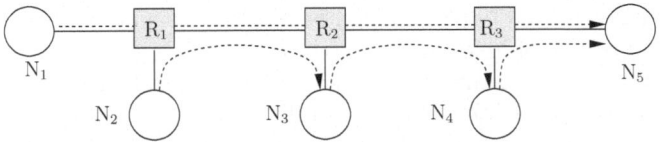

Fig. 5. Network with multiple bottlenecks to test the algorithm in the presence of background web traffic

accordingly. This confirms the need for an indirect way to detect packet losses anticipated in the introduction.

Finally, to validate our model in a more realistic scenario we used the network topology depicted in Fig. 5. In this setting, a single FAST-TCP flow carries data from N_1 to N_5 while thirty Reno flows carry web traffic from N_2 to N_3, N_3 to N_4 and N_4 to N_5. The loss pattern is shown in Fig. 6, where flows ids between 1 and 10 correspond to flows running from N_2 to N_3, ids in the range 11 to 20 belong to flows between N_3 and N_4, and the rest ids are to flows between N_4 and N_5. Note that, now, packet losses occur in three different bottlenecks and

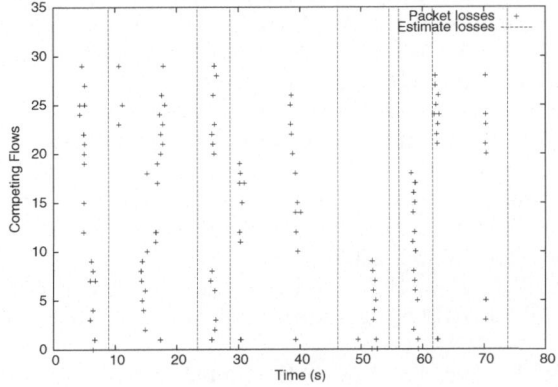

Fig. 6. Loss pattern in the presence of background web traffic in a network with multiple bottlenecks

not always simultaneously. Even so, the FAST-TCP flow from N_1 to N_5 is able to detect them, albeit with a lower accuracy, because simultaneous departures of HTTP sessions can be confused with the response to packet losses. Finally, some losses go undetected as the throughput of the FAST flow drops to 0 during heavy congestion episodes.

5   Conclusions

In the previous sections we have presented and described a new mechanism for the rapid passive detection of AIMD flows in a network path. Our proposal does not need of cooperation of any involved network agent and is able to accurately detect the occurrence of loses in just a few RTT.

The availability of this mechanism should help the creation of new hybrid congestion control protocols that exploit DCA advantages when in isolation, but can adapt to Reno-like protocols when needed, thus paving the way for the deployment of the first.

References

1. Floyd, S.: Highspeed TCP for large congestion windows. RFC 3649 (December 2003)
2. Rhee, I., Xu, L.: CUBIC: A new TCP-friendly high-speed TCP variant. In: PFLD-Net 2005 (February 2005)
3. Jain, R.: A delay-based approach for congestion avoidance in interconnected heterogeneous computer networks. SIGCOMM Comput. Commun. Rev. 19(5), 56–71 (1989)
4. Brakmo, L.S., O'Malley, S.W., Peterson, L.L.: TCP Vegas: New techniques for congestion detection and avoidance. SIGCOMM Comput. Commun. Rev. 24(4), 24–35 (1994)
5. Wei, D.X., Jin, C., Low, S.H., Hegde, S.: FAST TCP: Motivation, architecture, algorithms, performance. IEEE/ACM Trans. Netw. 14(6), 1246–1259 (2006)
6. Martin, J., Nilsson, A., Rhee, I.: Delay-based congestion avoidance for TCP. IEEE/ACM Trans. Netw. 11(3), 356–369 (2003)
7. Tang, A., Wei, D., Low, S.H.: Heterogeneous congestion control: Efficiency, fairness and design. In: IEEE International Conference on Network Protocols, Santa Barbara, CA, USA, 127–136 (November 2006)
8. King, R., Baraniuk, R., Riedi, R.: TCP-Africa: An adaptive and fair rapid increase rule for scalable TCP. In: Proceedings of the IEEE INFOCOM, Miami, FL, USA, vol. 3, pp. 1838–1848 (March 2005)
9. Wang, J., Wei, D.X., Low, H.S.: Modelling and stability of FAST TCP. In: Proceedings of the IEEE INFOCOM, April 2005, vol. 2, pp. 938–948, Pasadena, CA, USA (2005)
10. Choi, J.Y., Koo, K., Lee, J.S., Low, S.H.: Global stability of FAST TCP in single-link single-source network. In: 44th IEEE Conference on Decision and Control, Seville, Spain, pp. 1837–1841 (2005)
11. Choi, J.Y., Koo, K., Wei, D.X., Lee, J.S., Low, S.H.: Global exponential stability of FAST TCP. In: 45th IEEE Conference on Decision and Control, San Diego, CA, USA, December 2006, pp. 639–643 (2006)
12. Tan, L., Zhang, W., Yuan, C.: On parameter tuning for FAST TCP. IEEE Commun. Lett. 11(5), 458–460 (2007)
13. Mo, J., La, R.J., Anantharam, V., Walrand, J.: Analysis and comparison of TCP Reno and Vegas. In: Proceedings of the IEEE INFOCOM, New York, NY, USA, March 1999, vol. 3, pp. 1556–1563 (1999)
14. NS: ns Network Simulator (October 2005), http://www.isi.edu/nsman/ns/

Oscillations of the Sending Window in Compound TCP

Alberto Blanc[1], Denis Collange[1], and Konstantin Avrachenkov[2]

[1] Orange Labs, 905 rue Albert Einstein, 06921 Sophia Antipolis, France
[2] I.N.R.I.A. 2004 route des lucioles, 06902 Sophia Antipolis, France

Abstract. One of the key ideas of Compound TCP is to quickly increase the sending window, until full link utilization is detected, and then to keep it constant for a certain period of time. The actual Compound TCP algorithm does not hold the window constant but, instead, it makes it oscillate around the desired value. Using an analytical model and ns-2 simulations we study these oscillations on a Linux implementation of Compound TCP, in the case of a single connection with no cross traffic. Even in this simple case we show how these oscillations can behave in different ways depending on the bandwidth delay product. We also show how it is important to take into account, in the analytical model, that some implementation subtleties may introduce non-negligible differences in the behavior of the protocol.

1 Introduction

Tan et al. have introduced Compound TCP [5] to improve the performance of TCP on networks with large bandwidth delay products. One of the main ideas of this new high speed TCP variant is to quickly increase the sending window as long as the network is underutilized and then stabilizing it when a certain number of packets is buffered in the network. To achieve this goal the sender monitors the round trip time: as long as the network is underutilized (i.e. no packets are queued) the round trip time will not change. This corresponds to the case where the window is smaller than the bandwidth delay product. As the window is increased the sending rate will eventually surpass the capacity of the bottleneck link and the round trip time will start to increase. In particular the sender estimates the number of packets currently backlogged in the network. When this estimate is greater than or equal to a threshold γ the sender stops increasing its window. Figure 4 in [5], and some of the comments in that paper, give the impression that the window is then kept constant for a certain time.

While this might be a useful approximation in explaining and thinking about this new TCP version, it is easy to see, from equation (5) in [5], that the window will indeed oscillate during this phase. Figure 1 shows the evolution of the sending window from an ns-2 simulation. Clearly, at least in this case, the oscillations have a non-negligible amplitude. In the remainder of this paper we are going to analyze these oscillations in the case of a single Compound TCP connection with no competing traffic and no random losses (packets are dropped only when

E. Altman and A. Chaintreau (Eds.): NET-COOP 2008, LNCS 5425, pp. 95–102, 2009.

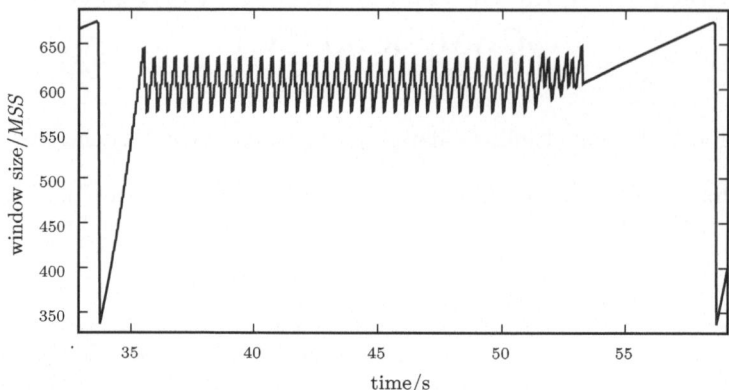

Fig. 1. The evolution of the sending window between two packet drops (ns-2 simulation, $\mu = 100\,\mathrm{Mb/s}$, $\tilde{\tau} = 69\,\mathrm{ms}$)

the buffer is full). While this is clearly the simplest possible scenario, even in this case, the oscillations have several possible patterns and, based on simulation results, their amplitude increases with the bandwidth delay product.

These oscillations may have many consequences on Compound TCP flows. Firstly, the packet loss probability in presence of other flows, using Compound TCP or not, will be increased during this phase. This may reduce the proportion of time during which the congestion window is larger than the bandwidth delay product, and then the efficiency of the protocol. Secondly, as these oscillations increase with the bandwidth delay product, the probability to saturate the bottleneck's buffer and will be higher and the efficiency will be lower in case of large round-trip times, as observed in [2]. Thirdly, these oscillations of the window, and, therefore of the buffer occupancy, may also degrade the performance of the network, increasing the delay, the jitter and the loss rate experienced by the other flows.

2   Evolution of the Compound TCP Window during the "Constant Window" Phase

During the "constant window" phase, every round trip time, the sender estimates the current backlog in the buffer of the bottleneck through the variable Δ (*diff* in [5]). An objective of Compound TCP is to keep Δ positive, to use the bottleneck at full speed, but small (close to γ), to minimize the buffer occupancy.

Let $w(t)$ be the window size at time t and let t_i be the end of the i-th round trip, when the window is increased from w_i to w_{i+1}. In order to make the $w(t)$ a left continuous function we set $w(t) = w_i$ if $t_{i-1} < t \leq t_i$. This way $w(t_i) = w_i$. Throughout this paper we will assume that w is expressed in terms of packets, or Maximum Segment Size (MSS). Note that $t_{i-1} - t_i$ is one round trip time so that the expression "every round trip time" refers to the time between two increments of the window. In the absence of loss, Compound TCP increments the window in the following way (see [5] for a complete description):

$$w_{i+1} = \begin{cases} w_i + \alpha w_i^k & \text{, if } \Delta_i < \gamma \\ w_i - \zeta\Delta_i + 1 & \text{, if } \Delta_i \geq \gamma \end{cases} \tag{1}$$

with

$$\Delta_i = w_i \left(1 - \frac{\tilde{\tau}}{\tau_i}\right). \tag{2}$$

Where $\tilde{\tau}$ is the smallest round trip time observed so far, and τ_i is the latest estimate of the round trip time. For the remainder of the paper we assume that $\zeta = 1$, and for all numerical example we use $\alpha = 1/8$, $\gamma = 30$, and $k = 3/4$, as suggested by the authors of CTCP [5]. While selecting these values is an interesting problem in itself it is outside the scope of this work (see [5] for more details).

Clearly as w increases so does Δ_i and, as the round-trip time is an increasing function of window, eventually $\Delta > \gamma$; and the window will be decreased, provided no packets are dropped in the meantime. Similarly, as smaller values of w imply smaller values of Δ, the window will be increased again. In other words one or more increasing phases are followed by one or more decreasing phases. We call a "cycle" the collection of increasing and decreasing phases starting with the first increasing phase and ending with the last decreasing phase. Figure 2 shows two possible cycles, the one on the left has 3 increasing phases and 1 decreasing phase, while the one on the right has 4 and 2, respectively. We will use two integers to classify cycles, with the first one representing the number of increments and the second one the number of decreasing phases. A 5:2 cycle, for example, has 5 increments and 2 reductions. In the figure dots and circles represents the value of the window *before* and *after* each increment, respectively.

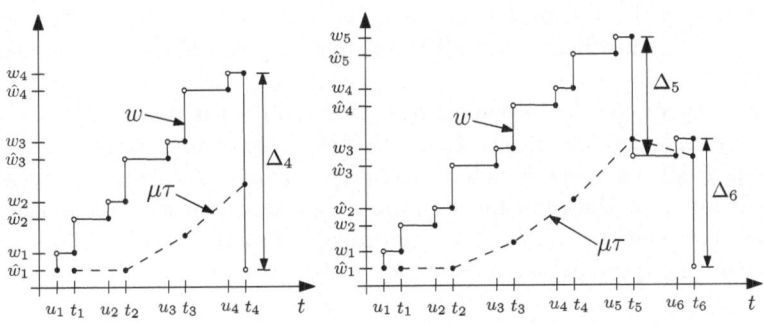

Fig. 2. The 3:1 and 4:2 cycles

The sending window of Compound TCP is the sum of two components that are incremented independently during each round trip time. The congestion component is incremented by one each round trip time, just like the window in TCP

Reno. While the delay component (w_d) is incremented, once per round trip time as well, as $w_d = w_d + \alpha w^k - 1$. The minus one compensates for the increase of the congestion component so that the total window grows as $w = w + \alpha w^k$. Similarly the plus one in (1) comes from the increment of the congestion window component, which happens also during the round trip time when the delay component is decreased. So that in a $m:n$ cycle there are m increments of the delay component and $m + n$ increments of the congestion component. In Figure 2 the smaller increments, between $\hat{w}_i$ and w_i, represent the increments of the congestion component and they take place at time u_i while the bigger increments are due to the delay component and take place at time t_i. Note that w_1 is the value *after* the initial increment of the congestion window, such increment represents the increment of the congestion window corresponding to the last reduction of the delay component at the end of the previous cycle.

2.1   Linux Implementation

In order to study the behavior of Compound TCP we ran simulations using ns-2 version 2.33 [4] with a Compound TCP implementation for Linux [1]. This implementation uses a slightly different formula to compute Δ_i: instead of w_i it uses w_{i-1} so that (2) is replaced by:

$$\Delta_i = w_{i-1}(1 - \tilde{\tau}/\tau_i). \tag{3}$$

Recall that w_i is the value of the window *before* the increment at time t_i so that w_{i-1} is the value of the window during $(t_{i-2}, t_{i-1}]$. While [5] is somewhat vague about the details of these computations, it is appropriate to use w_{i-1} instead of w_i. One way of thinking about Δ_i is that it tries to estimate the number of backlogged packets by comparing the current round trip time with the smallest round trip time observed so far. Given that the sender uses acknowledgments to measure the round trip time, any such sample corresponds to the round trip time experienced by the packet last acknowledged and such a packet was sent when $w = w_{i-1}$. In other words all the round trip samples are "one round trip time old."

Another aspect to take into account is that, while it is possible to find several equivalent expression for Δ_i, the Linux kernel does not use floating point operations. So that all the operations have to be approximated with integer ones. This introduces an error that can be minimized but that can lead to non-negligible differences between the implemented protocol and a theoretical model. In order to minimize the approximation error Δ_i is computed as:

$$2\Delta_i = 2w_{i-1} - \left\lfloor \frac{2w_{i-1}\tilde{\tau}}{\tau} \right\rfloor \tag{4}$$

and γ is multiplied by 2 whenever it is compared with Δ_i. (We use $\lfloor x \rfloor$ to represent the integer part of x.) Computing the window increment αw_n^k presents a similar problem. As $\alpha = 1/8$ and $k = 3/4$ (as suggested in [5]) the following formula is used:

$$\alpha w_n^k = \left\lfloor \frac{1}{2^3} \left\lfloor \frac{2^8 w_n}{\sqrt{2^{16} \sqrt{w_n}}} \right\rfloor \right\rfloor \tag{5}$$

where all the multiplications (and divisions) by a power of 2 are implemented as shift operations and the square root is implemented using the int_sqrt() function of the Linux kernel.

2.2   Modeling the Linux Implementation

For the case of a single connection with no other traffic $\tilde{\tau}$ is equal to the propagation delay, which we assume to be known. In this case it is also possible to express τ_i (the latest estimate of the round trip time) as a function of the window and the bottleneck capacity μ. In the Linux kernel the round trip times are measured in microseconds so that, even if integers are used, the precision is extremely high.

Assuming that the backlog is non-zero, we can compute the round trip time dividing the window by the bottleneck capacity so that:

$$\tau_i = \min \left[\left\lfloor \frac{w_{i-2} + 1}{\mu} \right\rfloor, \tilde{\tau} \right]. \tag{6}$$

This is is because the implementation in question uses the smallest round trip time sample among all the samples collected during the last round trip time, between t_{i-1} and t_i. (One comment in the source code explains that the choice of using the smallest round trip sample is to minimize the effect of delayed acknowledgments.) As the window, and therefore the round trip time, is an increasing function of time, the smallest value corresponds to the smallest time value; that is the beginning of the round trip. For example in Figure 2, at time t_4 the sender considers all the samples relative to the packets sent between times t_2 (excluded) and t_3 (included), whose acknowledgments were received between t_3 (excluded) and t_4 (included). At time t_2 the window was increased from w_2 to w_3, but, while this increment is instantaneous, the round trip time grows by smaller increments (more precisely by $1/\mu$) so that the smallest round trip sample *observed* between t_3 and t_4 corresponds to the packet that was sent when the window was $w_2 + 1$. Clearly τ_i cannot be smaller than the propagation delay, hence the minimum with $\tilde{\tau}$. This can indeed happen as w_i can be smaller than $\mu\tilde{\tau}$, either during the initial growing phase during the oscillation phase in the case of multiple reductions.

In [5] the authors say that the window should be updated "once per round trip time." The implementation we used accomplishes this as follows: whenever the window is updated the sequence number of the next segment to be transmitted (say n_i) is recorded. Once the corresponding acknowledgment arrives the window is updated another time. As a consequence, whenever the window is reduced because $\Delta \geq \gamma$ the next segment cannot be sent immediately (the window is smaller than the number of unacknowledged packets). The sender resumes transmission only after receiving acknowledgments for Δ packets. Due to this pause in the transmission the backlog experienced by packet n_i is smaller. When the corresponding acknowledgment arrives the round trip time (for this packet) is:

$$\tau_i = \min \left[\left\lfloor \frac{w_{i-1} - \Delta_{i-1}}{\mu} \right\rfloor, \tilde{\tau} \right]. \tag{7}$$

Where $w_{i-1} - \Delta_{i-1}$ is the window size after the reduction. Therefore, if at time t_{i-1} the window was reduced, t_i (i.e. the beginning of the new round trip time) corresponds with the arrival of the acknowledgment for packet n_i. As we have already mentioned, the implementation we have used takes the smallest of all the round trip samples collected between the last window update, at time t_{i-1} (excluded), and the current time t_i (included). This implies that, each time the window was reduced at t_{i-1}, the smallest round trip time sample is due to the last acknowledgment received. And the corresponding packet is the first one sent after the transmission pause.

Note that, provided the queue at the bottleneck link does not empty during the pause, when the sender resumes sending packets the backlog size will not change until the window is updated at time t_i. This implies that the value given by (7) is actually the "true" value of the round trip time between $t_{i-1} + \Delta_{i-1}/\mu$ and t_i , where Δ_{i-1}/μ is the duration of the transmission pause (it is the time needed to receive enough acknowledgments to compensate for the reduction of the window).

Given that we have defined a cycle as a series of increasing phases followed by one or more decreasing phase, the first phase of a cycle will always follow a window reduction so that, at t_1 (7) is used. In this case we also have that $w_m - \Delta_m = w_1 - 1$, where m is the last phase of the previous cycle. (Recall that we have defined w_1 as the value after the first increment of the congestion component.) And we can write $\tau_1 = (w_1 - 1)/\mu$. At time t_2 the smallest round trip samples received between t_1 and t_2 is again τ_1 because the packets whose acknowledgments arrive during $(t_1, t_2]$ were sent between during $(t_m + \Delta_m/\mu, t_1]$. As previously observed, all the packets sent during this time experience the same round trip time: $(w_m - \Delta_m)/\mu$.

In general, τ_i is given by (6) unless the window was reduced at time t_{i-1}, in which case (7) should be used, or if the window was reduced at time t_{i-2} and incremented at time t_{i-1}, in which case $\tau_i = \tau_{i-1}$. Using these formulas, together with those for Δ_i (3) and for the window increment (5) it is possible to model the evolution of the window. At the same time it is important to use the same integer approximations used in the Linux kernel: for example if we compute $w_4(w_1)$ (that is the value of the window after three increments with a starting value of w_1) using the same formulas used in the Linux Kernel or using floating point operations the difference between the two quantities is between 0 and 2. Where we use the formula αw_n^k to compute the window increment and compute Δ_i as $\Delta_i = w_i(1 - \tilde{\tau}/\tau)$ when we use floating point operations. Finally, Figure 3 compares Δ_5 using integer and floating point operations.

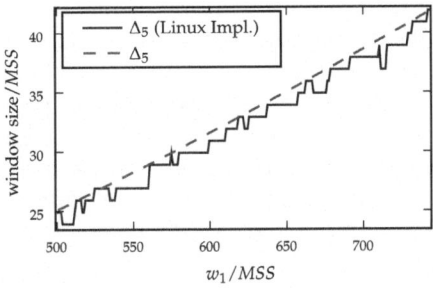

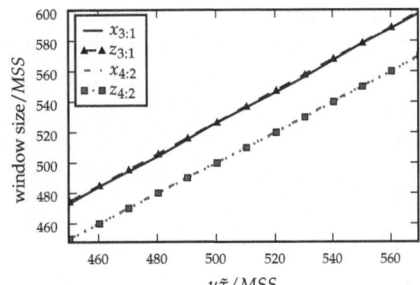

Fig. 3. Δ_5 using floating point and integer operations

Fig. 4. Solutions of the fixed point equation for the 3:1 and 4:2 cycles using floating point (z) and integer (x) operations

3   Fixed Points

Given that, as previously discussed, a series of increasing phases is always followed by one or more decreasing phases it is natural to ask if such oscillations follow a specific pattern and, above all, if they reach a steady state. It is sufficient to look at a few simulations to guess that oscillations do reach a steady state very quickly (after one or two cycles). Depending on the system parameters, we have observed 5 types of cycles: 3:1, 2:1, 5:2, 4:2 and 3:2. For each cycle type it is possible to find the steady state solution by numerically solving a fixed point equation. For the 3:1 cycle, for example, if w_1 is the value of the window at the beginning of each phase the value at the end of the cycle is $f_{3:1}(w_1) \triangleq w_4(w_1) - \Delta_4(w_1) + 1$ where $w_{n+1} = w_n + \alpha w_n^k$ and Δ_4 is given by (3). Note that the only independent variable is w_1 so that solving the fixed point equation $f_{3:1}(w_1) = w_1$ it is possible to find the steady state solution. The plus one takes into account the fact that, during each cycle, there are three increments of both window components and one increment (by one) of the congestion component combined with one reduction of the delay component. To be precise, here w_1 corresponds to the initial value of the total window after an increment by one of the congestion component (see Figure 2).

For each cycle $m:n$ it is possible to define the corresponding function $f_{m:1}(w_1)$ $= w_{m+1}(w_1) - \Delta_{m+1}(w_1) + 1$ if $n = 1$ and $f_{m:n} = w_{m+1}(w_1) - \Delta_{m+2}(w_1) + 2$ if $n = 2$. For the latter case the plus two compensates for the two increments of the congestion component corresponding to the two reductions of the delay component. Figure 4 shows the solution of $f_{3:1}$ and $f_{4:2}$ as a function of the bandwidth-delay product $(\mu\tilde{\tau})$. In this case the difference between using floating point and integer operations is not very significant, especially for the 4:2 case where the error is negligible. While Figure 4 shows only two cases, they are the most representative ones. The solutions for the 2:1 case are close to those of the 3:1 case. And the solutions for all the $m:2$ cases are almost identical. It is interesting to note how the solutions for the $m:2$ cases are very close (and in some cases equal) to $\mu\tilde{\tau}$. As $w_1 - 1$ is the minimum value of the window is

$w_1 = \mu \tilde{\tau}$ then $w_1 - 1 < \mu \tilde{\tau}$ which implies that the buffer will be empty for the first part of each oscillation. The simulations do confirm this, showing that the buffer will be empty for half a round trip time during each oscillation. This is caused by the increment of the congestion component, which usually takes place after half a round trip time after the last reduction. While half round trip time over a cycle of a few round trip times is not a big portion it does nonetheless lead to an under utilization of the bottleneck link during this "constant window" phase negating one of the main design ideas of Compound TCP.

4   Conclusions and Future Work

Contrary to what suggested by one of the figures in [5], we have shown how the Compound TCP window does oscillate during the "constant window" phase. These oscillations converge quickly to a cyclic behavior whose mode (number of increases:number of decreases) depends on the bandwidth delay product. Some implementation details on Linux have also an influence on the mode, and on the amplitude of these cycles, especially the discretization of the state variables, increments of the window and backlog estimates.

These oscillations may explain some of the inefficiencies observed for Compound TCP on some tests, especially when the round-trip times are large or the buffers small [3,2]. This phenomenon may also degrade the performance of the other simultaneous flows. We plan on further analyzing the influence of these oscillations on the performance of long lived Compound TCP connections and on other simultaneous flows. We would also like to understand the relationship between the parameters of the system and the type of cycles followed by the oscillations. While the simulations seem to indicate that this depends on the bandwidth delay product we do not yet know the exact relationship.

References

1. Andrew, L.: Compound TCP Linux module (April 2008), http://netlab.caltech.edu/lachlan/ctcp/
2. Baiocchi, A., Castellani, A., Vacirca, F.: YeAH-TCP: Yet Another Highspeed TCP. In: Proc. 5th Int. Workshop on Protocols for FAST Long-Distance Networks (March 2007)
3. Li, Y.-T.: Evaluation of TCP congestion control algorithms on the Windows Vista platform. Technical Report SLAC-TN-06-005, Stanford Linear Accelerator Center (June 2005)
4. McCanne, S., Floyd, S., et al.: ns network simulator, http://www.isi.edu/nsnam/ns/
5. Tan, K., Song, J., Zhang, Q., Sridharan, M.: A compound tcp approach for high-speed and long distance networks. In: INFOCOM 2006. Proc. 25th IEEE Int. Conf. on Computer Communications (2006)

Optimum Channel Allocation in OFDMA Multi-cell Systems

Andrea Abrardo[1], Paolo Detti[1], Gaia Nicosia[2],
Andrea Pacifici[3], and Mara Servilio[4]

[1] Università degli Studi di Siena,
Dipartimento di Ingegneria dell'Informazione, Siena, Italia
{abrardo,detti}@dii.unisi.it
[2] Università degli Studi Roma Tre,
Dipartimento di Informatica e Automazione, Roma, Italia
nicosia@dia.uniroma3.it
[3] Università degli Studi Tor Vergata,
Dipartimento di Ingegneria dell'Impresa, Roma, Italia
pacifici@disp.uniroma2.it
[4] Università degli Studi di L'Aquila,
Dipartimento di Informatica, via Vetoio, I-67010 Coppito, L'Aquila, Italia
servilio@di.univaq.it

Abstract. This paper addresses the problem of allocating users to radio resources (i.e., sub-carriers) in the downlink of an OFDMA cellular system. We consider a classical multi-cellular environment with a realistic interference model and a *margin adaptive* approach, i.e., we aim at minimizing total transmission power while maintaining a certain given rate for each user. We discuss computational complexity issues of the resulting model and present a heuristic approach that finds optima under suitable conditions, or "reasonably good" solutions in the general case. Computational experiences show that, in a comparison with a commercial state-of-the-art optimization solver, our algorithm is quite effective in terms of both infeasibilities and transmitted powers and extremely efficient in terms of CPU times.

Keywords: Radio resource allocation, network flow models, heuristic algorithms.

1 Introduction

IEEE 802.16 Air Interface Standard [6], which is the basis of the WiMAX technology, is the most recent solution for the provision of fixed broadband wireless services in a wide geographical scale and proved to be a real effective solution for the establishment of wireless metropolitan area networks (WirelessMAN). The most flexible physical layer implementation provided by the IEEE 802.16e standard is the OFDMA. It provides a sub-channelization structure with variable FFT (Fast Fourier Transform) sizes accommodating different channel bandwidths. The active sub-carriers are divided into subsets called sub-channels,

E. Altman and A. Chaintreau (Eds.): NET-COOP 2008, LNCS 5425, pp. 103–111, 2009.

being one of the basic units of resource allocation. In OFDMA systems, the concept of multi-user and single-user diversities are strictly connected. Indeed, for fixed or portable applications where the radio channels are slowly varying, an intrinsic advantage of OFDMA over other multiple access methods is its capability to exploit the multi-user diversity embedded in diverse frequency-selective channels [2,9]. Assigning the available sub carriers to the active users in an adaptive manner is a viable method to achieve multi-user diversity: the propagation channels are independent for each user and thus the sub carriers that are in a deep fade for one user may be good ones for another. Several papers have recently focused on the problem of optimum channel allocation of OFDMA cellular systems, and some of them have also considered the joint scheduling-allocation problem [10,12]. In general, resource allocation and scheduling tasks consist of either minimizing a cost measure (e.g. transmit power [5,10]) or maximizing a benefit (e.g. throughput [7,8,11]) while considering system hardware constraints, service specific quality of service (QoS) requirements and the overall system state. Assuming that the transmitter knows the instantaneous channel state of all users, significant performance improvements can be achieved in OFDMA if the sub-carrier, modulation, coding and power management is performed in an adaptive channel aware manner.

Most of the above studies have so far concentrated on the single cell scenario, where interference among users can be easily avoided by orthogonal assignments. In this context, it is shown that a high multi-user diversity gain can be obtained by means of cross layer approach, even in presence of a high degree of single user diversity. An interesting approach for considering more realistic interference model is considered in [1], where a classical cellular environment is studied. In [1], a realistic interference model is considered, where the ability of two nodes to communicate reliably depends not only on their distance but also on actual interference level which is produced by other users which are assigned the same resource, i.e., the same sub-carrier.

The paper is organized as follows. In Section 2, the problem is formally defined and its complexity is addressed. In Section 3, we proposed a heuristic algorithm for solving the radio resource allocation problem, based on a network flow model, which provides an optimal solution for a simplified version of the problem. Finally, computational results are presented and discussed in Section 4.

2   Problem Statement

In this section we describe the interference model used throughout the paper and give a formal statement of the optimization problem that we address.

We are given (i) a set of sub-carriers or radio resources $\{1, \ldots, m\}$; (ii) a set of cells $\{C_1, \ldots, C_k\}$, and (iii) for each cell C_h a set of n_h users $U_h = \{1, \ldots, n_h\}$. For each user i, we denote by $b(i)$ the cell i belongs to. Hence, $b(i) = C_h$ for all $i \in U_h$. If we set a certain target spectral efficiency η_i for user i, the transmission requirements correspond to a certain number of sub-carriers $r_i = R_i/\eta_i$, where R_i is the transmission rate required by user i, and η_i is set in a such a way that r_i is integer.

In general, users belonging to different cells can share the same sub-carrier, while interference phenomena do not allow two users in the same cell to transmit on the same sub-carrier. However, the power required for the transmission on a given sub-carrier increases with the number of users transmitting on that sub-carrier. More precisely, let S_j be the set of users which are assigned to (i.e., that are transmitting on) the same sub-carrier j. The transmission powers $p_i(j)$ requested by users in S_j on sub-carrier j satisfy the following system.

$$
\begin{aligned}
p_i(j) &= A_i(j) + \sum_{\substack{\ell \in S_j \\ \ell \neq i}} B_i^{b(\ell)}(j) p_\ell(j) \quad & i \in S_j \\
p_i(j) &\geq 0 & i \in S_j
\end{aligned}
\tag{1}
$$

where $A_i(j)$ and $B_i^{b(\ell)}(j)$ are given data taking into account the target signal-interference-ratio (SIR_i), the channel gain of user i on sub-carrier j ($G_i(j)$), and the channel gain between user i and the base station of cell $h \neq b(i)$ on sub-carrier j ($G_i^{b(h)}(j)$). More precisely, $A_i(j)$ is proportional to $SIR_i/G_i(j)$ and $B_i^{b(\ell)}(j) = SIR_i G_i^{b(\ell)}(j)/G_i(j)$. In the following, we refer to the quantities $A_i(j)$ and $\sum_{h \in S_j, h \neq i} B_i^{b(h)}(j) p_h(j)$ of System (1) as *fixed costs* and *variable costs*, respectively. Note that, System (1) may not have a feasible solution.

A *feasible radio resource allocation* consists in assigning sub-carriers to users in such a way that (i) for each user i, r_i radio resources are assigned to it, (ii) users in the same cell are not assigned to the same radio resource, (iii) given the set S_j of users assigned to radio resource j, System (1) has a feasible solution, for any sub-carrier $j = 1, \ldots, m$. Clearly, a necessary condition for a feasible allocation to exist is $m \geq \max_{h=1,\ldots,k} \sum_{i \in U_h} r_i$. The problem, that we call *Cellular Radio Resource Allocation* (RAP), consists of finding a feasible radio resource allocation that minimizes the total transmission power, i.e., the sum of the transmission powers required by all the users. Iteratively solving instances of RAP, on a radio-frame basis, may be used for dynamically assigning radio resources. In fact, practically a carrier is allocated to different users over time.

In Table 1, we summarize some results about the computational complexity for some special cases of the problem. All the details of the polynomial time algorithms and the NP-completeness proofs are reported in [4].

Table 1. Computational complexity results

Cells	Special Features	Complexity
2	cost as in (1)	open
2	identical resources	polynomial (Perfect Matching)
fixed k	indistinguishable users	polynomial (min cost flow)
3	identical resources	strongly NP-hard (from 3-dim axial assignment, [4])
-	identical $A_i(j)$ $\forall i, j$	strongly NP-hard (from 3-dim matching)
-	convex variable costs (Sect. 3)	polynomial (Min cost flow)
-	limited cell power	strongly NP-hard (from 3-Partition, [1])

In the first column of the table, the number of cells of the system is indicated, while the second column reports the special characteristics of the problems. In the problem with "identical resources", the fixed and variable costs do not depend on the particular resource the users are assigned to. The term "indistinguishable users" refers to the case in which the transmission power on a resource depends only on the number of users transmitting on it, and by the cell the users belong to. The case with convex variable costs is thoroughly addressed in Section 3. In the last row of the table, "Limited cell power" means that an upper bound on the transmission power of the users in the same cell is given. In this latter case, NP-hardness holds even if $n_h = 1$ for any cell $h = 1, \ldots, k$ and identical values of the $B_i^h(j)$'s.

3   A Network Flow Based Algorithm

In this section we propose a heuristic algorithm for solving RAP which is based on a minimum cost flow problem. To this purpose, we exploit a procedure that *exactly* solves a simplified version of RAP in which the transmission power of any sub-carrier j assumes a special structure.

3.1   A Simplified Model

Given a sub-carrier j and a set of users S assigned to it, suppose that the transmission power $T(j) = \sum_{i \in S} p_i(j)$ has the following special form:

$$T(j) = g(|S|) + \sum_{i \in S} A_i(j). \tag{2}$$

where $g(\cdot)$ is a convex function. Then $T(j)$ is comprised by the (usual) fixed cost part, which depends on the set S of users assigned to j, plus a *convex* variable cost which *only depends on the number* $|S|$ of users assigned to resource j. When the transmission power on resource j is expressed as in Equation (2) for all $j = 1, 2, \ldots, m$, we may find an optimal solution of RAP with the network flow model described hereafter.

Consider the marginal variable cost $\Delta_\ell(j) = g(\ell) - g(\ell - 1)$ corresponding to assigning an additional user to resource j when $\ell - 1 \geq 1$ users have been already assigned to it (let $\Delta_1(j) = 0$). Since g is convex, $\Delta_{\ell+1}(j) \geq \Delta_\ell(j) \geq 0$, for all $\ell \geq 1$.

On these grounds, we may define the following flow network $G = (V, A, c)$ illustrated in Figure 1. We distinguish four layers of nodes plus a demand node t. The first layer of nodes contains a supply node per each user. We denote the node associated to user i of cell C_h, as u_h^i. In the second and third layer, for any pair (j, h) associated to resource j and cell C_h, there is a node $v'_{j,h}$ for the second layer and $v''_{j,h}$ for the third layer . The fourth layer contains one node for any possible number of users ℓ assigned to resource j, for all $j \in R$. Since the maximum possible value for ℓ equals the number of cells k, we have mk nodes in this layer. We refer to the ℓ-th node corresponding to resource j as $w_{j,\ell}$. The arcs of the network are defined as follows.

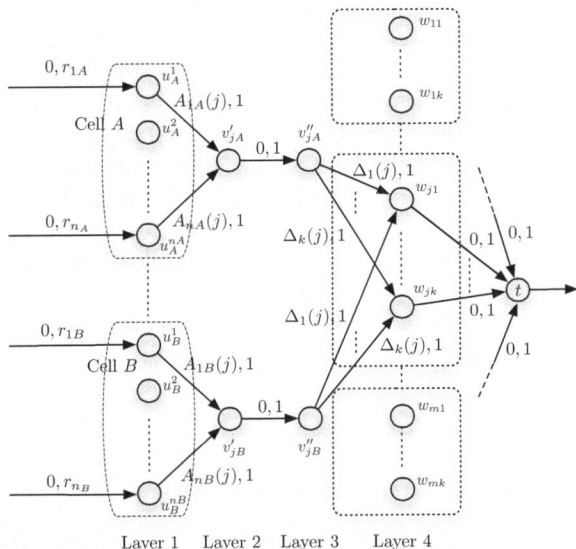

Layer 1   Layer 2   Layer 3   Layer 4

Fig. 1. The network flow model for the simplified RAP

1. For all $j \in R$, $h \in C$ and $i \in C_h$, arc $(u_h^i, v_{j,h}')$, with unit capacity and cost equal to $A_i(j)$, connects first and second layer nodes.
2. There is a unit capacity, null cost arc for each pair of nodes $v_{j,h}'$ to $v_{j,h}''$ between second and third layer.
3. For all $j \in R$, $C_h \in C$ and $1 \leq \ell \leq k$, unit capacity arc $(v_{j,h}'', w_{j,\ell})$ connect third and fourth layer nodes. The cost of such an arc is $\Delta_j(\ell)$.
4. All the arcs $(w_{j,h}, t)$ exist, with unit capacity and null cost, for all the nodes $w_{j,h}$ in the fourth layer.

The amount of commodity supplied by each node u_h^i in the first layer is r_i, the demand at node t is $\sum_i r_i$, while all other nodes are neither supply nor demand nodes. It is not hard to prove that there is a one-to-one correspondence between any integer feasible flow on G and a feasible solution of the instance of RAP. Therefore, the minimum cost flow in this supply-demand network provides an optimal solution to the simplified version of RAP.

3.2   A Heuristic for RAP

In this section, we propose a heuristic algorithm for RAP based on the network $G = (V, A, c)$ introduced in the previous section. Clearly, the cost structure of RAP does not allow to use the network model introduced above to determine an optimal solution. However, a feasible integer flow on G corresponds to an assignment of users to resources. The idea of the heuristic algorithm is that of approximating the actual costs of RAP with a cost structure satisfying Equation (2) and then applying the network flow model in order to obtain an *initial* assignment of users to the resources. Note that, such an initial assignment may not be feasible in the original problem, since System (1) may not be feasible for some set of users assigned to a resource. In most cases (and as long as a feasible solution of the original instance of RAP exists), possible infeasible assignments can be adjusted, via a local search procedure, to get a feasible solution. Even when feasible assignments exist, the procedure does not guarantee to find one of these solutions. However, as shown in Section 4, numerical experiments give an evidence that the procedure is indeed effective in this regard.

We proceed as follows, to approximate the costs in order to apply the model described in Section 3.1. In order to compute the variable costs part, we set $A_i(j) = \bar{A}(j)$ and $B_i(j) = \bar{B}(j)$ in System (1) for all $i \in U$ and $j \in R$, thus making the users indistinguishable. The linear system (1) becomes

$$p_i(j) = \bar{A}(j) + \bar{B}(j) \sum_{\substack{h \in S_j \\ h \neq i}} p_h(j) \quad i \in S_j. \tag{3}$$

After some algebra, the total transmission power required on resource j becomes

$$T_t'(j) = \sum_{i \in S_j} p_i(j) = \frac{t\bar{A}(j)}{1-(t-1)\bar{B}(j)} = t\bar{A}(j) + \frac{t\bar{A}(j)(t-1)\bar{B}(j)}{1-(t-1)\bar{B}(j)} \tag{4}$$

where $t = |S_j|$. Note that $T_t'(j)$ only depends on the number t of users assigned to j. According to the last expression of (4), $T_t'(j)$ can be decomposed into a fixed part, $t\bar{A}(j)$, and variable cost part, $g_j(t) = \frac{t\bar{A}(j)(t-1)\bar{B}(j)}{1-(t-1)\bar{B}(j)}$, as in (2). It is easy to show that $g(\cdot)$ is a convex function in t.

In our heuristic, we use the network described in Section 3.1, in which the costs of arcs connecting first and second layer nodes are the actual fixed costs coefficients $A_i(j)$, while the costs $\Delta_\ell(j)$ of arcs connecting the nodes of third and fourth layers are equal to $g_j(\ell) - g_j(\ell - 1)$. Observe that $g_j(t)$ could be negative for some value of t. In such cases, we say that it is *infeasible* to allocate t or more users to resource j, and we set $\Delta_\ell(j) = +\infty$, for all $\ell \geq t$. Observe that the flow between the second and third layer nodes defines a user-resource assignment and, as already mentioned, a feasible solution of the network flow problem may not correspond to a feasible solution for the original problem. Hence, the correspondence between the feasibility of the solutions of the two problems strictly depends on the choice of $\bar{A}(j)$ and $\bar{B}(j)$ parameters used to compute $g_j(\cdot)$.

We are now in the position to give a sketch of a heuristic, called NETWORK, for RAP. The procedure inputs a numerical instance of RAP and iteratively solves the network flow problem described above, as follows:

1. the parameters $\bar{A}(j)$ and $\bar{B}(j)$ are initially set to suitable values;
2. network $G = (V, A, c)$ is built and an optimal solution on G, i.e., an assignment $\mathcal{A}$ of users to resources, is found;
3. if an infeasibility is detected in $\mathcal{A}$, say on resource j, NETWORK carries out a local search in which, basically, the assignment of two users are exchanged: one user i is chosen in S_j and it is substituted by another user ℓ of the

same cell $b(i)$ (thus $\ell \notin S_j$). This search is performed for a given number of iterations or until all the infeasibilities are removed;

4. if $\mathcal{A}$ is feasible, some of the parameters $\bar{A}(j)$ and $\bar{B}(j)$ are updated, in order to improve the quality of the solution (i.e., the total transmission power). To this purpose, the fixed costs of those users requiring the highest transmission powers in the current assignment are increased.

Steps 1–4 are repeated for a limited number of iterations. Eventually, NET-WORK returns an assignment of all the users to the sub-carriers, if a feasible assignment has been found. If this is not the case, i.e., infeasibilities cannot be removed, the procedure returns an assignment of a subset of users (partial assignment). It is worthwhile to mention again that, given that the time required for collecting data and computation times are short enough, NETWORK may be used for dynamically assigning radio resources on a radio-frame basis.

4   Computational Experiments and Conclusions

In this section, computational results of the heuristic NETWORK on randomly generated instances are presented. The algorithm has been coded in standard C; CPLEX 9.1 has been used for solving the network flow problems. The performances of the heuristic have been compared with a truncated branch and bound algorithm that uses an Integer Linear Programming formulation solved with CPLEX 9.1 (most of the solutions are, in fact, optima or near-optimal solutions). In the instances, the number of cells is $K = 7$, the cell radius is 500 m, and the overall signal bandwidth is $B_{tot} = 5$ MHz. We assume a fixed throughput per cell equal to R_{tot} bit/s and that such a per-cell throughput is evenly shared among $|U_k| = n_k$ users, which are uniformly distributed in hexagonal cells of radius R. In all the instances, $m = 16$ sub-bands, each with a bandwidth of $B = 312.5$ kHz, and $n_k = 4$ users per cell have been considered. Hence, each user has a target throughput of $R_{tot}/4$ bit/s. As in [1], we assume that all users adopt the same transmission format, i.e., $\eta_i = \eta$ for all users on all sub-carriers. Since the rate per sub-carrier is $B\eta$, the condition to achieve the requested R_{tot} is that $\eta = \frac{R_{tot}}{16 \times B}$. Note that, in this case, each user is assigned a fixed number of sub-carriers $r_i = 4$. In all the instances $R_{tot} = B_{tot} \times \eta$. Hence, the bigger is η the bigger is the cell (fixed) target throughput. Three scenarios have been considered ($\eta = 2, 2.5, 3.0$) each corresponding to one hundred instances.

Table 2 illustrates a comparison of the results obtained by applying our NET-WORK heuristic (columns 2–4) and the CPLEX truncated branch and bound (columns 5–6). Average values of the following data are given: (i) solution, i.e., total transmission power, (ii) computation times in seconds, and (iii) a measure of rate loss. More precisely, column 3 reports the percentage of *resources with infeasible assignments* obtained by the NETWORK heuristic. Column 6 reports the *number of instances*, out of 100, in which the branch and bound cannot find any feasible solution within the time limit (these results have been presented in [1]). In fact, for the infeasible instances, the branch and bound algorithm cannot return an allocation for any user request. Note also that column 4 reports power

Table 2. Computational results

	Network			*Branch and Bound*		
η	Sol.value	Time	% Rate loss	Sol.value	Time	# Infeas.
2.0	67.59	0.31	0.0	48.56	201.68	0
2.5	251.18	0.64	0.07	101.92	268.78	2
3.0	710.92	1.51	0.46	215.58	339.56	18

values that are indeed computed over the set of instances for which a feasible solution exists. In most cases, NETWORK can find a feasible solution of reasonable quality in less than one second (which is less than one hundredth than the time required by the branch and bound.) This proves that an implementation of our algorithm on a dedicated processor is suitable for usage in an iterative scheme that solves the problem dynamically.

The power values (at least for the first scenario with $\eta = 2.0$) obtained by applying NETWORK, exceed those of the branch and bound by about 28% over all the 100 instances. However, if we compare the results, restricting our observations to the best (i.e., less congested) 85 and 70 instances, the gap decreases to 17% and 11%, respectively.

Possible future directions of research include (*i*) computational complexity characterization of "open" special cases of RAP; (*ii*) improvement of the proposed heuristic via the more effective local search scheme; (*iii*) design of *distributed* algorithms for RAP.

References

1. Abrardo, A., Alessio, A., Detti, P., Moretti, M.: Radio resource allocation problems for OFDMA cellular systems. Computers & Operations Research 36(5), 1572–1581 (2009)
2. Bender, P., Black, P., Grob, M., Padovani, R., Sindhushayana, N., Viterbi, A.: CDMA/HDR: a bandwidth-efficient high-speed wireless data service for nomadic users. IEEE Communications Magazine 38(7), 70–77 (2000)
3. Burkard, R.E., Rudolf, R., Woeginger, G.J.: Three-dimensional axial assignments problems with decomposable cost coefficients. Discrete Applied Mathematics 65, 123–139 (1996)
4. Detti, P., Nicosia, G., Pacifici, A., Servilio, M.: Optimal power control in OFDMA cellular networks. Tech. Report n. 2008-4, Dipartimento di Ingegneria dell'Informazione, University of Siena (2008)
5. Ergen, M., Coleri, S., Varaiya, P.: QoS aware adaptive resource allocation techniques. for fair scheduling in OFDMA based broadband wireless systems. IEEE Transactions on Broadcasting 49(4) (2003)
6. IEEE Standard for Local and Metropolitan Area Networks. Part 16: Air Interface for Fixed Broadband Wireless Access Systems, IEEE 802.16-2004 revision
7. Pietrzyk, S., Janssen, G.: Multiuser subcarrier allocation for QoS provision in OFDMA systems. In: IEEE Vehicular Technology Conference, VTC (2002)

8. Rhee, W., Cioffi, J.M.: Increase in capacity of multiuser OFDM system using dynamic subchannel allocation. In: Proc. IEEE Vehicular Technology Conference, VTC (2000)
9. Shakkottai, S., Rappaport, T.S., Karlsson, P.C.: Cross-Layer Design for Wireless Networks. IEEE Communications Magazine, 41(10) (2003)
10. Wong, C., Cheng, R., Lataief, K., Murch, R.: Multiuser OFDM with adaptive subcarrier, bit and power allocation. IEEE Journal on Selected Areas in Communications 17(10), 1747–1758 (1999)
11. Yin, H., Liu, H.: An efficient multiuser loading algorithm for OFDM based broadband wireless systems IEEE Globecom (2000)
12. Zhang, Y., Letaief, K.: Energy-efficient MAC-PHY resource management with guaranted QoS in wireless OFDM networks. In: Proc. IEEE ICC, Seoul, Korea (2005)

Transmission Power Control Game with SINR as Objective Function[*]

Eitan Altman[1], Konstantin Avrachenkov[1], and Andrey Garnaev[2]

[1] INRIA Sophia Antipolis, France
{altman,k.avrachenkov}@sophia.inria.fr
[2] St. Petersburg State University, Russia
agarnaev@rambler.ru

Abstract. We consider the transmission power control problem with SINR as objective function in the two scenarii: selfish and cooperative. We show that in the selfish (non-cooperative) scenario several Nash equilibria can arise. In particular, the game can take the form of the Hawk-Dove game, where the users can choose either conciliation or conflict fighting for shared sub-carriers. We fully characterize different types of Nash equilibria. In the cooperative scenario, we show that the parameter area where users employ pure strategies is essentially narrower than the area where users employ mixed strategies. Moreover, we identify an area where Nash equilibrium and Pareto equilibrium coincide. If one of the users has a large power resource (called a stronger user) for both scenarii and his rival has small power resource (weaker user) then the behaviour of the stronger user drastically changes in cooperative scenario compared to the selfish one. Namely, in the selfish scenario the stronger user squeezes the weaker one from the best channel meanwhile in the cooperative scenario he allows the weaker user to employ the best channel and himself applies a mixed strategy.

Keywords: Wireless networks, Power Control, Nash Equilibrium, Cooperation.

1 Introduction

We consider the transmission power control problem with SINR as objective function in the two scenarii: selfish and cooperative. In particular, in the selfish scenario we consider two users who try to send information through n resources. The strategy of user j ($j = 1, 2$) is $T^j = (T_1^j, \ldots, T_n^j)$ with $T_i^j \geq 0$ such that $\sum_{i=1}^{n} \pi_i T_i^j = \bar{T}^j$, where $\bar{T}^j > 0$. Here we assume that resource i has a "weight" of $\pi_i > 0$. The resources may correspond to capacity available at different time slots; we assume that there is a varying environment whose state changes among a finite set of states $i \in [1, n]$, according to some ergodic stochastic process with stationary distribution $\{\pi_i\}$. Either the resources may correspond to frequency

---

[*] The work was supported by EGIDE ECO-NET grant no.18933SL "Game Theory for Wireless Networks" and RFBR and NNSF Grant no.06-01-39005.

E. Altman and A. Chaintreau (Eds.): NET-COOP 2008, LNCS 5425, pp. 112–120, 2009.

bands (e.g. as in OFDM) where one should assign different power levels for different sub-carriers [9]. In that case we may take $\pi_i = 1/n$ for all i.

In the selfish (non-cooperative) scenario each user tries to maximize its average SINR:

$$v^j(T^1, T^2) = \sum_{i=1}^{n} \pi_i \frac{\alpha_i^j T_i^j}{N_i^0 + \alpha_i^{\hat{j}} T_i^{\hat{j}}},$$

where N_i^0 is the noise level and $\alpha_i^j > 0$ are fading channel gains of user j when the environment is in state i, and $\hat{j} = 3 - j$. We assume that all the fading channel gains α_i^j, the noise levels N_i^0, the total powers $\bar{T}^j$ are known to both users or they can be quickly inferred [4]. The authors of [1] have studied the case of incomplete information.

The SINR as an objective function in the power control game was also considered in [6]: all users have a single common channel and choose between several base stations. We note that in the regime of low SINR the present objective can serve as an approximation to the Shannon capacity. A central motivation to consider SINR as an objective function and not Shannon capacity, is that current technology for voice over wireless does not try to achieve Shannon capacity but rather uses given codecs that can adapt the transmission rate to the SINR; these turn out to adapt the rate in a way that is linear in the SINR over a wide range of throughput. The SINR has therefore been used very often to represent directly the throughput see [7,8]. The validity of this can be seen e.g. in [5, p. 151, 222, 239]. As we see from [5, Fig. 10.4, p. 222], the ratio between the throughput and the SINR is close to a constant throughout long range of bit rates. For example, between 16Kbps and 256Kbps, the maximum variation around the median value is less than 20%.

We finally note that with an SINR objective we are able to characterize fully cooperative and non-cooperative scenarii. In [2] with the Shannon capacity as an objective we have dealt only with the symmetric case. An interested reader can find more relevant literature on Gaussian Interference Game in [2,4]. In the present work we restrict ourselves to the case of two users. We believe that the two user case is good for illustration and many techniques of this paper apply to the case of more than two users which we leave as a topic for future research.

In the non-cooperative scenario we look for a NE (Nash Equilibrium), that is, we want to find such couple of strategies (T^{1*}, T^{2*}) that for any (T^1, T^2) the inequalities hold: $v^1(T^1, T^{2*}) \leq v^1(T^{1*}, T^{2*})$, $v^2(T^{1*}, T^2) \leq v^2(T^{1*}, T^{2*})$.

In the cooperative scenario, both users want to maximize $v^1 + v^2$. We shall show that in some parameter settings cooperative and non-cooperative strategies coincide.

It is worth to note that in [3] one of the results concerns a game which can be considered as a game with the SINR as object function in the jamming scenario the first user wants to maximize the objective function v_1 and the second user wants to minimize this objective function. For this game the jammer in his optimal behaviour tends to equalize the quality of the best sub-carriers to as low level as his power constraint allows.

We will assume that all channels differ by their quality for both users. Namely, we assume that for each user j ($j = 1, 2$) $\alpha_{i_1}^j / N_{i_1}^0 \neq \alpha_{i_2}^j / N_{i_2}^0$ for any $i_1 \neq i_2$. Without lost of generality we can assume that $\{a(j,1), a(j,2), \ldots, a(j,n)\}$ is a permutation of $\{1, 2, \ldots, n\}$ such that the sub-carriers are arranged in the following decreasing order by their quality: $\alpha_{a(j,1)}^j / N_{a(j,1)}^0 > \cdots > \alpha_{a(j,n)}^j / N_{a(j,n)}^0$.

2 Non-cooperative Scenario

2.1 The Best Sub-carriers Differs for the Users

Since v^i is linear in T^i the optimal strategy of user i can be nonnegative only for sub-carriers where the compound SINR is maximal. So, we have the following result describing the structure of the optimal strategies.

Theorem 1. (T^1, T^2) *is a Nash equilibrium if and only if there are non-negative* ω^1 *and* ω^2 *(which present the maximal compound SINR for the corresponding user) such that for* $j = 1, 2$ *and* $i \in [1, n]$:

$$T_i^j \geq 0 \text{ for } \frac{\alpha_i^j}{N_i^0 + \alpha_i^{\hat{j}} T_i^{\hat{j}}} = \omega^j \text{ and } T_i^j = 0 \text{ for } \frac{\alpha_i^j}{N_i^0 + \alpha_i^{\hat{j}} T_i^{\hat{j}}} < \omega^j.$$

The payoffs corresponding to these strategies are $(\omega^1 \bar{T}^1, \omega^2 \bar{T}^2)$.

The strategy when user j ($j = 1, 2$) transmits all the signal through just one sub-carrier (say, i) will be called a pure one and we will denote it by $\mathcal{T}_i^j$. So, the pure strategy is $\mathcal{T}_i^j = (T_1^j, \ldots, T_n^j)$ such that $T_k^j = \bar{T}^j$ for $k = i$ and $T_k^j = 0$ otherwise. The strategies when users employ more than one sub-carrier to transmit signal will be called mixed strategies.

If the best sub-carriers are different for users, namely $a(1,1) \neq a(2,1)$, then by Theorem 1 the NE has time sharing form, namely the following result holds.

Theorem 2. *If* $a(1,1) \neq a(2,1)$, *then the game has the unique NE and it is the pure one* $(\mathcal{T}_{a(1,1)}^1, \mathcal{T}_{a(2,1)}^2)$.

It is worth to note that if there are more than one best quality sub-carrier the NE is not unique at all since there is enough room for the users to share these best quality channels.

So, our assumption that all sub-carriers are different by its quality is quite reasonable. But as it will be shown in the next section even then if the best sub-carriers coincides a variety of NE is possible.

2.2 The Best First Sub-carriers Coincides for Users

Let the first best sub-carriers coincides for users, i.e. $a(1,1) = a(2,1) = a(1)$. Then some variety of cases arise depending on the quality of the second best sub-carriers and the power of signal the users can apply.

The SINRs for the best sub-carrier with the induced noise are big enough for both users

If the SINRs for the best sub-carrier with the induced noise are greater than the SINRs with the natural noise for the second best sub-carriers for both users, then each of them could manage to transmit all the signal just through the best sub-carrier. The next theorem follows straighforward from Theorem 1 and supplies the formulas describing the corresponding conditions on SINRs.

Theorem 3. *Let the following inequalities hold*

$$\frac{\alpha_{a(1)}^{j}}{N_{a(1)}^{0} + \alpha_{a(1)}^{j}\bar{T}^{j}} \geq \frac{\alpha_{a(j,2)}^{j}}{N_{a(j,2)}^{0}} \; for \; j = 1, 2. \tag{1}$$

Then there is the unique NE and it is the pure one $(T_{a(1)}^{1}, T_{a(1)}^{2})$.

Note that inequalities (1) can be rewritten in the following equivalent form

$$A^{j} \geq \bar{X}^{j} \; for \; j = 1, 2, \tag{2}$$

where $A^{\hat{j}} = \alpha_{a(1)}^{j} N_{a(j,2)}^{0} / \alpha_{a(j,2)}^{j} - N_{a(1)}^{0}$ and $\bar{X}^{j} = \alpha_{a(1)}^{j}\bar{T}^{j}$.

The SINR for the best sub-carrier with the induced noise is big enough only for one user

If the SINR for the best sub-carrier with the induced noise is greater than the SINR with the natural noise for the second best sub-carriers only for one user (say, user $\hat{j}$), then the other user (user j) could threaten to jam the best sub-carrier and user $\hat{j}$ has nothing to threaten back. This makes him to withdraw and to use the second best sub-carrier. The next theorem follows straightforward from Theorem 1 and supplies the formulas describing the corresponding relations between SINRs.

Theorem 4. *Let* $A^{j} > \bar{X}^{j}$, $A^{\hat{j}} < \bar{X}^{\hat{j}}$ *hold either for* $j = 1$ *or* $j = 2$: *Then the game has the unique NE and it is the pure one* (T^{1}, T^{2}) *where* $T^{j} = T_{a(j,2)}^{j}$ *and* $T^{\hat{j}} = T_{a(1)}^{\hat{j}}$.

Thus, the weaker user is squeezed out from the common best sub-carrier by the threat from the stronger user.

The SINRs for the best sub-carrier with the induced noise are small for both users

If the SINRs for the best sub-carrier with the induced noise are less than the SINRs with the natural noise for the second best sub-carriers for the both users, then each user could threaten to jam the best sub-carrier for his opponent. The situation becomes very competitive and Hawk-Dove type strategies are possible.

The next theorem follows straightforward from Theorem 1 and supplies the formulas describing the corresponding relations between SINRs and Hawk-Dove type strategies.

Theorem 5. *Let $A^j < \bar{X}^j$ hold for $j = 1, 2$. Then there are two pure NE equilibria $(\mathcal{T}^1_{a(1)}, \mathcal{T}^2_{a(2,2)})$ and $(\mathcal{T}^1_{a(1,2)}, \mathcal{T}^2_{a(1)})$.*

In the first pure NE user 1 presents an aggressive (Hawk) player always fighting for the best quality sub-carrier meanwhile user 2 is a withdrawing (Dove) player escaping any fighting. In the second pure NE the users exchange the roles. In Figure 1 possible pure NE are pointed out depending on power signals they have to transmit.

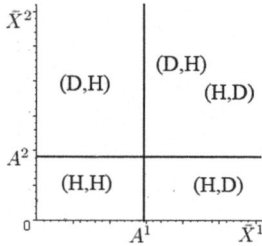

Fig. 1. Pure NE strategies

Besides the two pure NE a mixed NE can take place. Let us consider the case when the second best sub-carriers are different for users. Then by Theorem 1 we can see that in the mixed equilibrium both users at first take care about the opponent by equalizing quality of the opponent's two best by quality sub-carriers. It allows users to transmit all the rest of the signal through the second best sub-carriers. Of course, this strategy assumes some cooperation between the users.

Theorem 6. *Let $A^j < \bar{X}^j$ hold for $j = 1, 2$ hold and $a(1, 2) \neq a(2, 2)$, so the second best sub-carriers are different for users. Then there is the unique mixed NE, using only two first by quality sub-carriers, given by the following strategies which equalize the first and second by quality sub-carriers and transmit the rest power through the second sub-carriers:*

$$
T^j_i = \begin{cases}
\frac{\alpha^{\hat{j}}_{a(1)}}{\alpha^j_{a(1)}}\left(1/\omega^{\hat{j}} - N^0_{a(1)}/\alpha^{\hat{j}}_{a(1)}\right), & \text{for } i = a(1), \\
\bar{T}^j - T^j_{a(1)}, & \text{for } i = a(j, 2), \\
0, & \text{otherwise,}
\end{cases}
\tag{3}
$$

where $\omega^j = \alpha^j_{a(j,2)}/N^0_{a(j,2)}$, $j = 1, 2$.

2.3   The K Best Sub-carriers Coincide

In this subsection we consider situation when the K $(K \leq n)$ best sub-carriers coincide for users, i.e. $a(1,i) = a(2,i) = a(i)$, $i \in [1,K]$ and $a(1, K+1) \neq a(2, K+1)$ if $K < n$. Theorem 6 can be generalized as follows:

Theorem 7. *Let $k^1 = k^2 < K$ where $k^j (j = 1,2)$ are integers such that*

$$H^{\hat{j}}(\alpha^j_{a(j,k^j)}/N^0_{k^j}) < \bar{T}^{\hat{j}} \leq H^{\hat{j}}(\alpha^j_{a(j,k^j+1)}/N^0_{k^j+1}),$$

where $\alpha^j_{a(j,n+1)}/N^0_{n+1} = 0$ and

$$H^{\hat{j}}(\omega) = \sum_{i=1}^{n} \frac{\alpha^j_{a(j,i)}}{\alpha^{\hat{j}}_{a(j,i)}} \left[1/\omega^j - N^0_{a(j,i)}/\alpha^j_{a(j,i)}\right]_+.$$

Then the NE is given by the strategies (T^1, T^2) where

$$T^{\hat{j}}_i = \frac{\alpha^j_{a(i)}}{\alpha^{\hat{j}}_{a(i)}} \left[1/\omega^j - N^0_{a(i)}/\alpha^j_{a(i)}\right]_+, \quad i \in [1,n]$$

for $j = 1,2$, and ω^j is the unique root of the equation $H^{\hat{j}}(\omega^j) = \bar{T}^{\hat{j}}$.

Thus, if the powers of signal to transmit are approximately equal then the equilibrium mixed strategies have water-filling structure.

Theorem 8. *Let $K = k^1 = k^2$ and $a(1, K+1) \neq a(2, K+1)$. Then if the power of signals allows to equalize the first $K+1$ channels, so*

$$H^{\hat{j}}\left(\alpha^j_{a(j,K+1)}/N^0_{a(j,K+1)}\right) < \bar{T}^{\hat{j}}$$

then the NE is given by the strategies (T^1, T^2) where

$$T^j_i = \begin{cases} \dfrac{\alpha^{\hat{j}}_{a(i)}}{\alpha^j_{a(i)}} \left[1/\omega^{\hat{j}} - N^0_{a(i)}/\alpha^{\hat{j}}_{a(i)}\right]_+ & \text{for } i = a(1), \ldots, a(k), \\ \bar{T}^j - H^j(\omega^{\hat{j}}) & \text{for } i = a(j, K+1), \\ 0 & \text{otherwise,} \end{cases}$$

and where $\omega^j = \alpha^j_{a(j,K+1)}/N^0_{a(j,K+1)}$, $j = 1,2$.

3   A Cooperative Scenario

In this section we assume that the users cooperate and they want to maximize their joint payoff $v = v^1 + v^2$. Of course, if the best sub-carriers are different for users, then, time sharing strategies $T^1_{a(1,1)}$, $T^2_{a(2,1)}$ are the optimal ones. Next, let us consider the case when the best sub-carriers coincide for the users, namely $a(1,1) = a(2,1) = a(1)$. The next theorem supplies the complete solution of this problem where the second best sub-carriers are different, namely $a(1,2) \neq a(2,2)$.

Theorem 9. *Let $a(1,1) = a(2,1) = a(1)$ and $a(1,2) \neq a(2,2)$. (i) If*

$$F^j(\bar{X}^1, \bar{X}^2) \geq \bar{\alpha}^j \text{ for } j = 1, 2, \tag{4}$$

where

$$F^j(\xi^1, \xi^2) = \alpha^j_{a(1)} \left(\frac{1}{N^0_{a(1)} + \xi^{\hat{j}}} - \frac{\xi^{\hat{j}}}{(N^0_{a(1)} + \xi^{\hat{j}})^2} \right), \bar{\alpha}^j = \frac{\alpha^j_{a(j,2)}}{N^0_{a(j,2)}} \tag{5}$$

then $(T^1_{a(1)}, T^2_{a(1)})$ is optimal.
(ii) If

$$F^1(\bar{X}^1, \bar{X}^2) < \bar{\alpha}^1 \text{ and } F^2(\bar{X}^1, \bar{X}^2) > \bar{\alpha}^2 \tag{6}$$

then (T^1, T^2) is optimal, where $T^2 = T^2_{a(1)}$, $T^1 = (t, \bar{T}^1 - t, 0, \ldots, 0)$, $t = \max\{0, T\}$ and T is the maximal root of the equation:

$$\alpha^1_{a(1)} \left(\frac{1}{N^0_{a(1)} + \alpha^1_{a(1)}\bar{T}^2} - \frac{\alpha^1_{a(1)}\bar{T}^2}{(N^0_{a(1)} + \alpha^2_{a(1)}T)^2} \right) = \bar{\alpha}^1. \tag{7}$$

(iii) The case $F^1(\bar{X}^1, \bar{X}^2) > \bar{\alpha}^1$ and $F^2(\bar{X}^1, \bar{X}^2) < \bar{\alpha}^2$ is symmetric to (ii).
(iv) If

$$F^j(\bar{X}^1, \bar{X}^2) < \bar{\alpha}^j \text{ for } j = 1, 2 \tag{8}$$

then (T^1, T^2) is optimal, where $T^j = (x^j/\alpha^j_{a(1)}, \bar{T}^j - x^j/\alpha^j_{a(1)}, 0, \ldots, 0)$ and (x^1, x^2) is the positive solution of the system $F^j(x^1, x^2) = \bar{\alpha}^j, j = 1, 2$.

This theorem allows us to depict (see Figure 2) in coordinate system $(\bar{X}^1, \bar{X}^2)$ how NE and cooperative equilibria differ depending on the power signal users can apply. The area $\{(\bar{X}^1, \bar{X}^2) : X^j \leq A^j, j = 1, 2)$, where the pure equilibrium $(T^1_{a(1)}, T^2_{a(1)})$ is NE, is essentially larger than the area $\{(\bar{X}^1, \bar{X}^2) : F^j(\bar{X}^1, \bar{X}^2) \geq \bar{\alpha}^j\}$ where it is also the cooperative equilibrium. Meanwhile the area $\{(\bar{X}^1, \bar{X}^2) : F^j(\bar{X}^1, \bar{X}^2) > \bar{\alpha}^j\}$ where in cooperative behavior the users prefer to employ mixed strategies include the corresponding area for NE as well as a part of the area where in NE a user is squeezed from the best sub-carrier. Thus, in the cooperative scenario in comparison with the non-cooperative scenario the area where users prefer to employ pure strategies is essentially narrower.

The conditions that $F^j(\bar{X}^1, \bar{X}^2) > \bar{\alpha}^j\}$ and $\bar{X}^j < \bar{A}^j$ can be interpreted as if user j has a small power resource for cooperative and selfish scenarii, respectively, meanwhile the conditions that $F^j(\bar{X}^1, \bar{X}^2) < \bar{\alpha}^j\}$ and $\bar{X}^j > \bar{A}^j$ can be interpreted as if user j has a large power resource. If one of the users has a large power resource (called a stronger user) for both scenarii meanwhile his rival has small power resource (weaker user) then behaviour of the stronger user drastically changes in the cooperative scenario compared to the selfish one. Namely, in selfish one the stronger user squeezes the weaker one from the best channel meanwhile in the cooperative scenario he allows the weaker user to employ the best channel and himself applies a mixed strategy. Finally note that the lines on Figure 2 can be also interpreted as switching lines for the cost of anarchy.

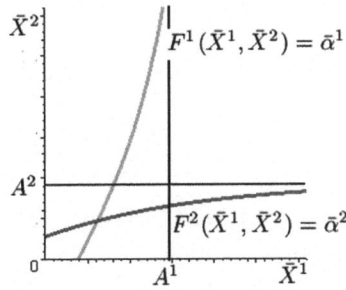

Fig. 2. Cooperative and Nash equilibria

4   Appendix

Proof of Theorem 6. Since $a(1,2) \neq a(1,2)$ the users do not have colluded interests in the second by quality sub-carriers which are different for them. Also, $T^1_{a(1)} > 0$ and $T^2_{a(1)} > 0$ since the users apply mixed strategies. Thus, by Theorem 1, $\omega^1 = \alpha^1_{a(1,2)}/N^0_{a(1,2)}$, $\omega^2 = \alpha^2_{a(2,2)}/N^0_{a(2,2)}$, and

$$\frac{\alpha^1_{a(1)}}{N^0_{a(1)} + \alpha^2_{a(1)} T^2_{a(1)}} = \omega^1, \qquad \frac{\alpha^2_{a(1)}}{N^0_{a(1)} + \alpha^1_{a(1)} T^1_{a(1)}} = \omega^2.$$

Solving these equations for T^1 and T^2 implies (3).

Proof of Theorem 9. Since $a(1,2) \neq a(2,2)$ and $a(1,1) = a(2,1)$ the optimal strategy T^j, $j = 1,2$ has to be of the form

$$T^j_i = \begin{cases} t^j & \text{for } i = a(1), \\ \bar{T}^j - t^j & \text{for } i = a(j,2), \\ 0 & \text{otherwise,} \end{cases}$$

where $t^j \in [0, \bar{T}^j]$ and the optimal t^1 and t^2 are the ones which maximize the following function

$$v(t^1, t^2) = \frac{\alpha^1_{a(1)} t^1}{N^0_{a(1)} + \alpha^2_{a(1)} t^2} + \frac{\alpha^2_{a(1)} t^2}{N^0_{a(1)} + \alpha^1_{a(1)} t^1} + \sum_{i=1}^{2} \frac{\alpha^i_{a(i,2)}}{N^0_{a(i,2)}} (\bar{T}^i - t^i).$$

To find the optimal t^1 and t^2, we need to calculate the derivatives of v with respect to t^1 and t^2:

$$v_j(t^1, t^2) := \frac{\partial v}{\partial t^j} = \frac{\alpha^j_{a(1)}}{N^0_{a(1)} + \alpha^{t^j}_{a(1)} t^j} - \frac{\alpha^1_{a(1)} \alpha^2_{a(1)} t^{\hat{j}}}{(N^0_{a(1)} + \alpha^{t^j}_{a(1)} t^{\hat{j}})^2} - \frac{\alpha^j_{a(j,2)}}{N^0_{a(j,2)}}, \ j = 1,2.$$

It is clear that

$$v_j(t^1, t^2) = \tilde{v}_j(\xi^1, \xi^2) := F^j(\xi^1, \xi^2) - \bar{\alpha}^j,$$

where $\xi^j = \alpha^j_{a(1)} t^j$ and $\xi^j \in [0, \tilde{T}^j]$ with $\tilde{T}^j = \alpha^j_{a(1)} \bar{T}^j$, $j = 1,2$.

$\tilde{v}_j$ has the following properties:

$$\tilde{v}_j \text{ is increasing in } \xi^j \text{ and decreasing in } \xi^{\hat{j}}, \tag{9}$$

$$\tilde{v}_1(t,0) = \frac{\alpha_{a(1)}^1}{N_{a(1)}^0} - \frac{\alpha_{a(1,2)}^1}{N_{a(1,2)}^0} > 0, \ \tilde{v}_2(0,t) = \frac{\alpha_{a(1)}^2}{N_{a(1)}^0} - \frac{\alpha_{a(2,2)}^2}{N_{a(2,2)}^0} > 0 \text{ for } t \geq 0. \tag{10}$$

By (10) the points (0,0), (0,ξ), (ξ,0) cannot be the maximal ones.

In Figure 3 we depict signs of $\tilde{v}_1$ and $\tilde{v}_2$ in the coordinate system (ξ^1, ξ^2). Namely, the region (+,+) presents the set $\{(\xi^1, \xi^2) : \tilde{v}_1(\xi^1, \xi^2) > 0, \tilde{v}_2(\xi^1, \xi^2) > 0\}$, the region (−,−) is $\{(\xi^1, \xi^2) : \tilde{v}_1(\xi^1, \xi^2) < 0, \tilde{v}_2(\xi^1, \xi^2) < 0\}$ and so on. If $(\tilde{T}^1, \tilde{T}^2) \in (-,-)$ which corresponds to (5) then the maximum is at $(\tilde{T}^1, \tilde{T}^2)$ and (i) is proved. If $(\tilde{T}^1, \tilde{T}^2) \in (+,+)$ which corresponds to (8) then the maximum is at (ξ^1, ξ^2) which is the unique positive solution of the system $\tilde{v}_i(\xi^1, \xi^2) = 0, i = 1, 2$ and (iv) is proved. Similarly, the cases $(\tilde{T}^1, \tilde{T}^2) \in (+,-)$ and $(\tilde{T}^1, \tilde{T}^2) \in (-,+)$ correspond to (8) respectively and are investigated similarly.

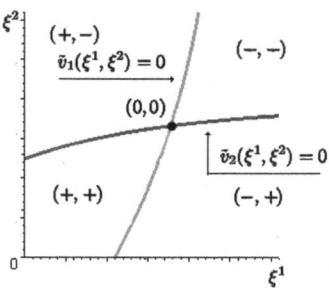

Fig. 3. Signs of $\tilde{v}_1$ and $\tilde{v}_2$

References

1. Adlakha, S., Johari, R., Goldsmith, A.: Competition in Wireless Systems via Bayesian Interference Games. Arxiv.org, no.0709.0516
2. Altman, E., Avrachenkov, K., Garnaev, A.: Closed form solutions for symmetric water filling games. In: Proc. of IEEE INFOCOM (2008)
3. Altman, E., Avrachenkov, K., Garnaev, A.: Fair resource allocation in wireless networks in the presence of a jammer. In: Proc. of ValueTools 2008 (2008)
4. Etkin, R., Parekh, A., Tse, D.: Spectrum Sharing in Unlicensed Bands. IEEE J. on Selec. Areas Comm. 25(3), 517–528 (2007)
5. Holma, H., Toskala, A.: WCDMA for UMTS
6. Ji, H., Huang, C.-Y.: Non-cooperative uplink power control in cellular radio systems. Wireless Networks 4, 233–240 (1998)
7. Kim, S.L., Rosberg, Z., Zander, J.: Combined power control and transmission selection in cellular networks. In: Proceedings of IEEE Vehicular Technology Conference (Fall 1999)
8. Koo, I., Ahn, J., Lee, H.A., Kim, K.: Analysis of Erlang capacity for the multimedia DS-CDMA systems. IEICE Trans. Fundamentals E82-A(5), 849–855 (1999)
9. Tse, D., Viswanath, P.: Fundamentals of Wireless Communication. Cambridge University Press, Cambridge (2005)

Battery State-Dependent Access Control in Solar-Powered Broadband Wireless Networks*

Hamidou Tembine[1], Eitan Altman[2], Rachid El-Azouzi[1], and Yezekael Hayel[1]

[1] LIA/CERI, University of Avignon. 339, chemin des Meinajaries Agroparc BP 1228
84911 Avignon Cedex 9 France
[2] INRIA B.P 93, 2004 Route des Lucioles 06902 Sophia Antipolis Cedex France

Abstract. This paper studies both power control and multiple access control in solar-powered broadband wireless networks. We assume that the mobiles use power storage element, such as rechargeable Solar-powered batteries, in order to have energy available for later use. By modeling the energy-level of Solar-powered batteries as a stochastic process, we study noncooperative interactions within large population of mobiles that interfere with each other through many local interactions. Each local interaction involves a random number of mobiles. The actions taken by a mobile determine not only the immediate payoff but also the state transition probabilities of its battery. We define and characterize the evolutionary Stable Strategies (ESS) of the stochastic evolutionary game.

Keywords: Power Control, Access Control, Evolutionary games, Markov Decision Processes.

1 Introduction

Power control and multiple access control in wireless networks have become an important research area. Since today's technology cannot provide batteries with small weight and large energy capacity, design of tools and protocols for efficient power control are needed in order to improve the energy capacity for small batteries. For a comprehensive survey of recent results on power control in wireless networks, an interested reader can consult e.g., [4] and the references therein.

Environmental energy is becoming a feasible alternative for many low-power systems, such as wireless sensor and mesh networks. However, this provides an ubpredictable and limited amount of energy over time. The power storage elements, such as rechargeable batteries or super-capacitors, become very useful to increase the system lifetime and the system availability. In particular, solar power is made possible with the use of Photovoltaic cells. Comprised of several layers of material, these cells are able to produce electrical power from exposure to sunlight. Since in many geographic areas, nice weather is not guaranteed and is

* This work was partially supported by the INRIA ARC Program POPEYE and the European Project BIONETS.

E. Altman and A. Chaintreau (Eds.): NET-COOP 2008, LNCS 5425, pp. 121–129, 2009.

unpredictable, the nodes should be able to recover from blackout periods caused by the unavailability of energy. In [3], a stochastic model for a solar powered wireless sensor/mesh networks is used to analyze the following QoS measures for several stochastic policies proposed: the average battery capacity, the sleeping probability and the average delay.

In this paper, we consider an evolutionary game approach with dynamic rechargeable battery depending on the weather (solar energy). Evolutionary games have been used to model the evolution of population sizes as a result of competition between them that occurs through many local interactions. Central in evolutionary games is the concept of Evolutionary Stable Strategy (ESS), which is a distribution of (deterministic or mixed) actions such that if used, the population is immune against penetration of mutations. This notion is stronger than that of Nash equilibrium as ESS is robust against a deviation of a whole fraction of the population where as the Nash equilibrium is defined with respect to possible deviations of a single player.

We make use of a recent dynamic version of stochastic evolutionary games, which we call "Markov Decision Evolutionary Games" (MDEG)[1], in solar-powered broadband wireless networks. There are many local interactions among individuals belonging to large populations of mobiles. The result of the interaction between mobiles depends on their current individual state. ¿From time to time the individual state of a mobile varies. The actioni choice of mobiles involved in a local interactions as well as their individual states determine the not only the result of the interactions but also the transition probabilities to the other possible individual states. Each individual is thus faced with an Markov Decision Process (MDP) in which it maximizes the expected average cost criterion. Each individual knows only the state of its own MDP, and does not know the state of the other mobiles it interacts with.

The destination of some transmission occasionally may receive simultaneously a transmission from another terminal which results in a collision. It is assumed however that even when packets collide, one of the packets can be received correctly if transmitted at a higher power. As state of the MDP of a user we take its energy level. The immediate fitness (rewards) is the number of successful transmissions. By allowing the mobiles to be equipped with rechargeable solar powered batteries, the mobiles may have infinite life time and the criteria that is maximizing is the limit average Cesaco-type payoff.

The remainder of this paper is organized as follows. In Sections 2 and 3, we present the description and notation used in our modeling approach and study the special case of three pure strategies in battery-state dependent power and access control. Section 5 concludes the paper.

2   The Model

Consider the following setting of evolutionary games: there is a large populations of mobiles; each mobile has a finite number of transmission power level available. There are many local interactions at the same time. At each slot, some of the terminals have to take a decision on their transmission power based on their own battery state. At the lowest state of the battery no power is available and the mobile has to wait the time to have good weather for regain some energy. Each player has its individual states set $S = \{0, 1, 2, \ldots, n\}$. Each mobile of the population has a finite action set in each state $s : A(s)$. We assume that there are a random number of interacting mobiles in each local interaction. At time t, each mobile knows its own state s_t and selects an action $a_t \in A(s_t)$. The mobile receives some payoff $r(s_t, a_t, \alpha_t)$ where α_t is the composition of the population at time t (the $j-$th element of $\alpha_t^{s_t}$ represents the fraction of mobiles choosing the action j) and the state of mobile goes to the state s_{t+1} with the probability $q(s_{t+1}|s_t, a_t)$. Each individual is thus faced with an MDP in which it maximizes the expected average cost criterion. Each individual knows only the state of its own MDP, and does not know the state of the other players it interacts with. The transition probabilities of a player's MDP are only controlled by that player. Let $\Delta(S)$ be the $(|S| - 1)$-dimensional simplex of the Euclidean space $\mathbb{R}^{|S|}$. The set of all action profiles at all states is given by $All = \{(s, a), \ s \in S, \ a \in A(s)\}$, then $q : \quad All \times \Theta \rightarrow \Delta(S)$ is a transition rule between the states where $\Theta = \prod_{s \in S} \Delta(A(s))$, $\Delta(A(s))$ is the set of probability distribution on $A(s)$. The vector $[q(s_{t+1} = 0|s_t, a_t), \ldots, q(s_{t+1} = n|s_t, a_t)]$ satisfies $\forall \ j \in S$, $q(s_{t+1} = j|s_t, a_t) \geq 0$, $\sum_{j \in S} q(s_{t+1} = j|s_t, a_t) = 1$. A state s is absorbing state if $q(s|s, a) = 1$, $\forall \ a \in A(s), \forall \alpha \in \Theta$. We examine the limit average Cesaro-type payoff given a population profile σ and an individual trajectory u,

$$F_n(u, \sigma) = \liminf_{T \longrightarrow +\infty} \mathbb{E}_{u, \sigma} \left(\frac{1}{T} \sum_{t=1}^{T} r(s_t, a_t, \alpha_t) \right)$$

where $\mathbb{E}_{u,\sigma,n}$ denotes the expectation over the probability measure $\mathbb{P}_{u,\sigma,n}$ induced by u, σ on the set of histories endowed with the product $\sigma-$algebra (initial state of the battery is n). Define further

- The subset $\mathcal{U}_S$ of stationary policies; a stationary policy u is a policy in which the probability to choose an action a depends only on the current state s; it is denoted by $u(a|s)$.
- The subset $\mathcal{U}_D \subset \mathcal{U}_S$ of pure or deterministic stationary policies $\mathcal{U}_D$. A policy of this type can be viewed as a function from the states to the actions.
- The set $\mathcal{U}_M$ of mixed strategies: A mixed strategy is identified with a probability γ over the set of pure stationary strategies. It can be considered as first choosing a pure stationary policy u with probability $\gamma(u)$ and then keeping choosing forever the actions according to u. A general mixed strategy is a mixture of behaviorial strategy.

Occupation Measure. Often we encounter the notion of individual states in evolutionary games; but usually the population size at a particular state is fixed. In our case the choices of actions of an individual determine the fraction of time it would spend at each state. Hence the fraction of the whole population that will be at a given state may depend on the distribution of strategies in the population. In order to model this dependence we first need to introduce the expected average frequency $f_{n,u}(s)$ that an individual spends at a given state s when it follows a strategy u and its initial state at time 1 is n. Moreover, we define $f_{n,u}(s,a)$ the expected average frequency during which it is at state s and it chooses action a, $f_{u,n} = \lim_{T_o \infty} \frac{1}{T} \sum_{t=1}^{T} Q^t(u)$ where $Q(u) := [q(s'|s,a,u)]_{s',s,a}$.

We now define the equilibrium concept in the context of evolutionary games. A behaviorial stationary strategy $u = (u_1, \ldots, u_n)$ with $u_i = (x_i, y_i)$ is an evolutionary stable strategy (ESS) if for all strategy mut such that $f_{mut} \neq f_u$ there exists $\epsilon_{mut} > 0$ such that

$$\sum_{s,a}(f_u(s,a) - f_{mut}(s,a))r(s,a,\alpha^\epsilon) > 0 \text{ where } \alpha^\epsilon = (1-\epsilon)\alpha(u) + \epsilon\alpha(mut).$$

Interpretation: Suppose that, initially, the population profile is $\alpha(u)$. Now suppose that a small group of mutants enters the population playing according to a different profile $\alpha(mut)$. If we call $\epsilon \in (0,1)$ the size of the subpopulation of mutants after normalization, then the population profile after mutation will be $\epsilon\alpha(mut) + (1-\epsilon)\alpha(u)$. After mutation, the average payoff of non-mutants will be given by $\sum_{s,a} f_u(s,a)r(s,a,\alpha^\epsilon)$. Analogously, the average payoff a mutant is $\sum_{s,a} f_{mut}(s,a)r(s,a,\alpha^\epsilon)$. That is, u is ESS if, after mutation, non-mutants are more successful than mutants. In other words, mutants cannot invade the population and will eventually get extinct. In case where the payoff function r is linear in last variable α (it is not the case in this model) then this definition is equivalent to $\alpha(u)$ is a strict symmetric Nash equilibrium or $\alpha(u)$ is a strictly better response than $\alpha(mut)$ given that the others mobiles mut. Note that if α is not linear, we can have $\alpha(\epsilon mut + (1-\epsilon)u) \neq \epsilon\alpha(mut) + (1-\epsilon)\alpha(u)$.

3   Stochastic Modeling of the Energy Levels of Solar-Powered Battery

In this subsection, we suppose that a battery has $n+1$ energy states $S = \{0, 1, \ldots, n\}$. The state 0 corresponds to the state *Empty* and the state n is the *Full* state of the battery. The other states $1, \ldots, n-1$ are intermediary states of the battery. We associate with each mobile a Markov Decision Process (MDP) which represents the transition probabilities between energy levels. Let X_t^i be the energy level of battery at time t. Given a stationary policy σ and a strategy of all the population $\alpha = (\alpha_t)_{t \geq 1}$, the transition probability of the energy level of battery is is described by the (first order, time-homogeneous) Markov process (X_t) where the transition probability law q which is given by

$$q_{sas'} = \begin{cases} 1 - R_{\gamma,s}(a) - Q_{\gamma,s}(a) & \text{if } s' = s-1 \\ R_{\gamma,s}(a) & \text{if } s' = s+1 \\ Q_{\gamma,s}(a) & \text{if } s' = s \\ 0 & \text{otherwise} \end{cases}, \forall 1 \leq s \leq n-1, \forall a_s \in A(s),$$

$$q_{nas'} = \begin{cases} 1 - Q_{\gamma,n}(a) & \text{if } s' = n-1 \\ Q_{\gamma,n}(a) & \text{if } s' = n \\ 0 & \text{otherwise} \end{cases}, \quad q_{0as'} = \begin{cases} \gamma & \text{if } s' = 1 \\ (1-\gamma) & \text{if } s' = 0 \\ 0 & \text{otherwise} \end{cases},$$

where $\gamma \longmapsto R_{\gamma,s}(a) \in [0,1]$ is an increasing function $\forall s, a$ with $R_{0,s}(a) = 0$, $0 \leq R_{\gamma,s}(a) + Q_{\gamma,s}(a) \leq 1$. The factor γ represents the probability to have a "good weather". If γ is zero, the state 0 is unique absorbing sate and expected lifetime of the battery is finite. For $\gamma \neq 0$, the Markov chain is irreducible i.e., there is only one class, that is, all states communicate with each other for stationary policy.

3.1 Battery-State Dependent Access Control in Solar-Powered System

This subsection studies is a generalization of the *random access game* with unknown number of mobiles, finite state, and three strategies. The channel is ideal for transmission and all errors are due to collision. A mobile can transmit a packet using a power level among three available levels: transmit with high power P_h, transmit with low power P_l or does not transmit 0. We consider a general capture model where a packet transmitted by a mobile is received successfully when if and only if that mobile uses a transmission power which is greater than the power used by the others transmitters at that time slot. Given a population profile $\alpha = (x, y, 1 - x - y)$ with $0 \leq x, 0 \leq y, x + y \leq 1$, the expected probability to have a successful transmission:

(i) When the mobile chooses P_l, the reward is the probability that the others mobiles choose zero i.e no others mobiles transmit, i.e.,

$$r(s, P_l, \alpha) = \sum_{k \geq 0} \mathbb{P}(K = k)(1 - x - y)^k = \mathbb{E}((1 - x - y)^K) = G_K(1 - x - y)$$

where G_K is the generating function of K.

(ii) When the mobile chooses P_h, the reward is the probability that no other mobiles transmit with the high power i.e.,

$$r(s, P_h, \alpha) = \mathbb{E}_K \sum_{l=0}^{K} \binom{K}{l}(1 - x - y)^l y^{K-l} = \mathbb{E}_K(1 - x)^K = G_K(1 - x). \quad (1)$$

where $\binom{K}{l}$ is the binomial coefficient of K and l.

(iii) When the terminal chooses 0, the reward is zero, i.e., $r(s, 0, \alpha) = 0$

Since there are three strategies P_h, P_l and 0, the population aggressiveness can be described by a process $(x_t, y_t)_{t \geq 1}$ where x_t (resp. y_t) is the fraction of the population using the high power level (low power) at time t. However, for each state $s \neq 0$, the action space becomes $M := \{(x, y), x \geq 0, y \geq 0, x + y \leq 1\}$. A stationary policy of an user is a map $\beta : S \longrightarrow M$. The expected reward of a user when its battery is at the state $s \neq 0$ is then given by

$$\tilde{r}(s, \beta, \alpha) = x'_s r(s, P_l, \alpha) + y'_s r(s, P_l, \alpha),$$

where $\beta(s) = (x'_s, y'_s) \ \forall s \in S$.

4   Computing Equilibria and ESS

4.1   Nash Equilibria and Pareto Optimality

In this subsection, we study the existence and uniqueness of Nash equilibrium in different scenarios:

Two mobiles $K = \delta_2$. The interaction in each slot and each non-empty individual state is described in the following tabular. Mobile 1 chooses a row, mobile 2 chooses a column. The payoff of mobile 1 (resp. mobile 2) is the first (resp. the second) component of the vector payoff.

1\2	"0"	P_l	P_h
"0"	$(0,0)$	$(0,1)^\diamond$	$(0,1)^{\diamond\star}$
P_l	$(1,0)^\diamond$	$(0,0)$	$(0,1)^{\diamond\star}$
P_h	$(1,0)^{\diamond\star}$	$(1,0)^{\diamond\star}$	$(0,0)^\star$

The Nash equilibria are represented by $\star$ and Pareto optimal allocation[1] are represented by $\diamond$.

Three mobiles $K = \delta_3$. The instantaneous reward is presented in the matrix game. Mobile 1 chooses a row, mobile 2 chooses a column and mobile 3 chooses an array M_i.

$$
M_1 := \begin{array}{|c|c|c|}
\hline
(0,0,0) & (0,1,0)^\diamond & (0,1,0)^\star \\
\hline
(1,0,0)^\diamond & (0,0,0) & (0,1,0)^\star \\
\hline
(1,0,0)^{\diamond\star} & (1,0,0)^{\diamond\star} & (0,0,0)^\star \\
\hline
\end{array}
\quad
M_2 := \begin{array}{|c|c|c|}
\hline
(0,0,1)^\diamond & (0,0,0) & (0,1,0)^{\diamond\star} \\
\hline
(0,0,0) & (0,0,0) & (0,1,0)^{\diamond\star} \\
\hline
(1,0,0)^{\diamond\star} & (1,0,0)^{\diamond\star} & (0,0,0)^\star \\
\hline
\end{array}
$$

$$
M_3 := \begin{array}{|c|c|c|}
\hline
(0,0,1)^{\diamond\star} & (0,0,1)^{\diamond\star} & (0,0,0)^\star \\
\hline
(0,0,1)^{\diamond\star} & (0,0,1)^{\diamond\star} & (0,0,0)^\star \\
\hline
(0,0,0)^\star & (0,0,0)^\star & (0,0,0)^\star \\
\hline
\end{array}
$$

Proposition 4.2. *In any state $s \neq 0$, the one-shot local interaction between $p \geq 2$ mobiles has an infinite (Nash) equilibria, $\binom{p}{1}2^{p-1} = p2^{p-1}$ of them are Pareto optimal, and a unique symmetric (Nash) which is the strategy h (independently of the state).*

Note that this one-shot game has a unique evolutionary stable strategy (see [2]).

Proof. The Nash equilibria and Pareto optimality of the local interaction in state $s \neq 0$ can be described as follows:

(i) **Symmetric Equilibrium:** It is clear that $r(s, P_h, \alpha) \geq r(s, P_l, \alpha) \geq r(s, 0, \alpha)$, $\forall \alpha$ i.e P_h weakly dominates P_l which weakly dominates 0. Hence, the strategy P_h is an equilibrium. Moreover the best reply to the population profile $\alpha = (x, y, 1 - x - y)$ is to play P_h if $x \neq 1$, and to play any strategy

---

[1] An allocation of payoffs is Pareto optimal or Pareto efficient if there is no other allocation that makes every node at least as well off and at least one node strictly better off.

$z \in [0,1]$ if $x = 1$. Thus, $(P_h, P_h, \ldots, P_h)$ is the unique symmetric equilibrium. At the equilibrium $(P_h, P_h, \ldots, P_h)$, the reward of each mobile is zero. Thus, $(P_h, P_h, \ldots, P_h)$ is not Pareto optimal because the allocation obtained at $(P_h, P_l, \ldots, P_l)$ or $(P_h, 0, 0, \ldots, 0)$ Pareto dominates zero.

(ii) **Pure Equilibria:** Fix a mobile m which uses the action P_h. Then any action profile of the others mobiles $b^{-m} \in \in \{0, P_l, P_h\}^{p-1}$ leads to a Nash equilibrium (no mobile can improve its probability of success by deviating unilaterally). In particular, if k $(0 \leq k \leq p-1)$ of the $p-1$ mobiles choose P_l and the $p-k-1$ others use 0 then no mobile can improve its probability of success by deviating unilaterally. The mobile m has exactly $\sum_{k=0}^{p-1} \binom{p-1}{k} = 2^{p-1}$ pure equilibria in which he/she has successful transmission. By changing the role of m, we get $\binom{p}{1} 2^{p-1} = p 2^{p-1}$ pure equilibria with successful transmission. All these pure equilibria are Pareto optimal: if only one terminal uses the high power P_h and the others mobiles use 0 or P_l, then the mobile with the high power P_h gets the payoff 1 and the others gets the payoff 0.

(iii) **Mixed Equilibria:** Any situation where at least one of the mobiles use the strategy P_h , and other mobiles use an arbitrary mixed strategies, gives a mixed Nash equilibria. The allocation of payoff obtained in these mixed strategy are not Pareto optimal if at least one mobile chooses the strategy P_h with positive probability.

In the figure 1, we plot the best response (power level) of a mobile for a given trajectory of a profile population, with $P_h = 10$ and $P_l = 2$. We observe that the best response is is to use a low power when the population is very aggressive $(x = 1)$, and to use the high power in any state $s \neq 0$ when the population is less aggressive $x < 1$. In the figure 2, we plot the sojourn time of a mobile to stay in a state as function of its stationary strategy $\beta = (x, y)$ with $s \neq 0, Q_s(0) = 0.95, Q_s(P_l) = 0.65$ and $Q_s(P_h) = 0.25$. We observe that the sojourn time is decreasing function in aggressiveness of that mobile.

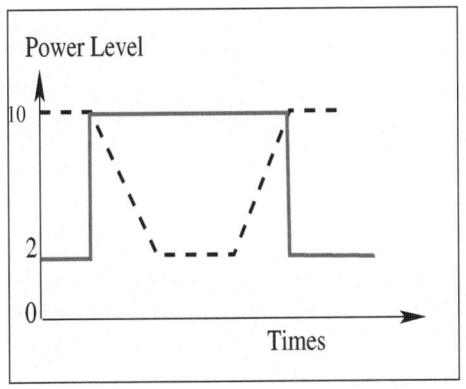
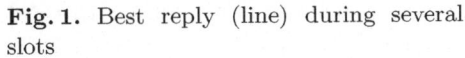
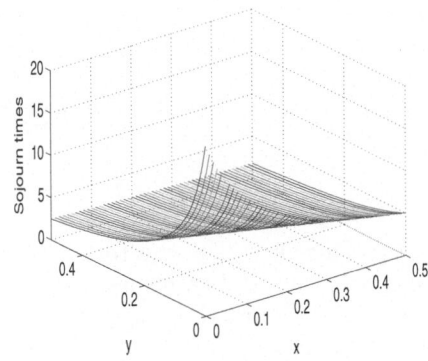

Fig. 1. Best reply (line) during several slots

Fig. 2. Sojourn time

4.3 Evolutionary Stable Strategy

Under the condition $\gamma > 0$, we have the following result [1]:

Proposition 4.4. *The payoff function has a representation in term of occupation measure, i.e., $F_{n,u}(\sigma) = \sum_{s,a} f_u(s,a)r(s,a,\alpha)$ where $\alpha = (x,y)$ is the profile of population under stationary strategy σ.*

Proposition 4.5. *The fully aggressive strategy $\bar{P}_h$ which consists to transmit with the maximum power (P_h) in each state $s \neq 0$ is an Evolutionarily Stable Strategies.*

Proof. Let mut such that $\alpha'(mut) \neq 1$. For $u = \bar{P}_h$ the time average payoff is $F_u(\alpha^\epsilon)$ $= G_K \left(\epsilon(1 - \alpha'_{\bar{P}_h}(mut)) \right) \sum_{s=1}^n f_{\bar{P}_h}(P_h)$, and $F_{mut}(\alpha^\epsilon) = \sum_{s=1}^n f_{mut}(s, mut_s) \times$ $\left[mut_{P_h} G_K(\epsilon(1 - \alpha'_{\bar{P}_h}(mut))) + mut_{P_l} G_K(\epsilon(1 - \alpha'_{\bar{P}_h}(mut) - \alpha'_{P_l}(mut))) \right]$. Since the generating is increasing, $G_K(\epsilon(1 - \alpha'_{\bar{P}_h}(mut) - \alpha'_{P_l}(mut))) \leq G_K(\epsilon(1 - \alpha'_{\bar{P}_h}(mut))$ with equality if and only $\alpha'_{P_l}(mut) = 0$. We deduce that $F_{mut}(\alpha^\epsilon)$ is strictly greater $F_{mut}(\alpha^\epsilon)$ for all mut such that $\alpha'_{\bar{P}_h}(mut) < 1$.

4.6 Power Control in Clouded Weather

We assume now that the solar-powered system is in a clouded weather ($\gamma = 0$, "no sunlight") during a long period (clouded sky, raining time or due to the season). Power storage elements, such as supercapacitors (but finite in practice), in order to have energy available for later use has been proposed. Because of limited capacity (hence energy of the battery), the aggressive terminals (which use the high power) will be rapidly in the state 0 which becomes an absorbing state. The terminals with empty battery need an alternative solution such as external recharge or to buy a new battery. So there is an additional cost to survive in this situation. If the maximum lifetime of the battery (for example using the power "0") is finite then the power control in clouded weather can be modeled by the stochastic evolutionary game with total successful transmission before to reach to the absorbing state 0.

Lemma 4.6.1 *The total expected successful transmission during the lifetime of the battery under the strategy u is given by*

$$\frac{1}{1-\gamma} \times 0 + \sum_{s=1}^n \frac{u_s(P_l)G_K(1 - \alpha_{P_l} - \alpha_{P_h}) + u_s(P_h)G_K(1 - \alpha_{P_h})}{1 - Q_{0,s}(u_s)}$$

where $Q_{0,s}(u_s) := u_s(P_h)Q_{0,s}(P_h) + u_s(P_l)Q_{0,s}(P_l)$.

Proof. This says that the total reward total is the sum over the states of the expected successful transmission times the expected sojourn times spent in this state. The sojourn time in state s under the policy u_s is $\frac{1}{1-Q_{0,s}(u_s)}$. This completes the proof.

The following Proposition holds in the stochastic evolutionary game with total reward.

Proposition 4.7. *(i) The strategy "stay quiet" (plays "0" in each state) cannot be an ESS. (ii) A necessary condition for the full aggressive strategy to be an ESS is $G_K(0) = P(K = 0) = 0$. Moreover if $G_K(1 - \alpha_{P_h}) \sum_{s=1}^{n} \frac{1}{1-Q_{0,s}(P_h)} >$ $\max_{mut \in \{P_h, P_l\}^n} \left\{ \sum_{s=1}^{n} \frac{r(s,mut_s,\alpha)}{1-Q_{0,s}(mut_s)} \right\}$ then $\bar{P}_h$ is an ESS. (iii)Similarly, if G_K $(1 - \alpha_{P_h}) \sum_{s=1}^{n} \frac{1}{1-Q_{0,s}(P_l)} > \max_{mut \in \{P_h, P_l\}^n} \left\{ \sum_{s=1}^{n} \frac{r(s,mut_s,\alpha)}{1-Q_{0,s}(mut_s)} \right\}$ then $\bar{P}_l$ is an ESS.*

Proof. The strategy "stay quiet" (play "0" in each state) cannot be an ESS because it is best reply to itself. Hence, the strategy *not transmit* can be invaded by mutations. If $\bar{P}_h$ is an ESS then $G_K(0) \sum_{s=1}^{n} \frac{1}{1-Q_{0,s}(P_h)}$ must be greater than $G_K(0) \sum_{s=1}^{n} \frac{1}{1-Q_{0,s}(P_l)}$. Since the power consumption is greater with P_h than P_l ($P_h > P_l$), one has, $\sum_{s=1}^{n} \frac{1}{1-Q_{0,s}(P_h)} < \sum_{s=1}^{n} \frac{1}{1-Q_{0,s}(P_l)}$. This implies that $G_K(0) = P(K = 0) = 0$. The other results are immediate by best response conditions.

5 Concluding Remarks

Thanks to the renewable energy techniques, designing autonomous mobile terminal and consumer embedded electronics that exploit the energy coming from the environment is becoming a feasible option. However, the design of such devices requires the careful selection of the components, such as power consumption and the energy storage elements, according to the working environment and the features of the application. In this paper we have investigated power control interaction based on stochastic modeling of the remaining energy of the battery for each user in each local interaction. We have showed existence of equilibria and conditions for evolutionary stable strategies.

References

1. Altman, E., Hayel, Y.: A Stochastic Evolutionary Game of Energy Management in a Distributed Aloha Network. In: Proc. of INFOCOM 2008 (2008)
2. Altman, E., Elazouzi, R., Hayel, Y., Tembine, H.: Evolutionary Power Control Games in Wireless Networks. In: Proc. 7th IFIP Networking 2008 (2008)
3. Niyato, D., Hossain, E., Fallahi, A.: Sleep and wakeup strategies in solar-powered wireless sensor/mesh networks: Performance analysis and optimization. IEEE Transactions on Mobile Computing 6(2), 221–236 (2007)
4. Meshkati, F., Poor, H., Schwartz, S.: Energy-Efficient Resource Allocation in Wireless Networks. IEEE Signal Processing Magazine 58 (2007)
5. Susu, A.E., Acquaviva, A., Atienza, D., De Micheli, G.: Stochastic Modeling and Analysis for Environmentally Powered Wireless Sensor Nodes. In: Wiopt 2008 (2008)
6. Tembine, H., Altman, E., El-Azouzi, R., Hayel, Y.: Evolutionary games with random number of interacting players applied to access control. In: WiOpt 2008 (2008)

Author Index